People, Plants and Protected Areas

PEOPLE AND PLANTS CONSERVATION MANUALS

Manual Series Editor
Martin Walters

Manual Series Originator
Alan Hamilton

People and Plants is a joint initiative of WWF, the United Nations Educational, Scientific and Cultural Organization (UNESCO) and the Royal Botanic Gardens, Kew.

Other titles in the series

Applied Ethnobotany: People, Wild Plant Use and Conservation
Anthony B Cunningham

Plant Invaders: The Threat to Natural Ecosystems (reissue)
Quentin C B Cronk and Janice L Fuller

Forthcoming titles in the series

Biodiversity and Traditional Knowledge: Equitable Partnerships in Practice
Sarah A Laird (ed)

Ethnobotany: A Methods Manual 2nd edition
Gary J Martin

The Management and Marketing of Non-Timber Forest Products: Certification as a Tool to Promote Sustainability
Patricia Shanley, Sarah A Laird, Alan Pierce and Abraham Guillén (eds)

Uncovering the Hidden Harvest: Valuation Methods for Woodland and Forest Resources
Bruce M Campbell and Martin K Luckert (eds)

People, Plants and Protected Areas

A Guide to *In Situ* Management

John Tuxill and Gary Paul Nabhan

WWF

UNESCO

ROYAL BOTANIC GARDENS KEW

earthscan
from Routledge

First published by Earthscan in the UK and USA in 2001

For a full list of publications please contact:
Earthscan
2 Park Square, Milton Park, Abingdon, Oxon OX14 4RN
Simultaneously published in the USA and Canada by Earthscan
711 Third Avenue, New York, NY 10017

First issued in hardback 2017

Earthscan is an imprint of the Taylor & Francis Group, an informa business

Originally published as *Plants and Protected Areas* by Stanley Thornes in 1998

A catalogue record for this book is available from the British Library

ISBN 13: 978-1-138-40711-4 (hbk)
ISBN 13: 978-1-85383-782-1 (pbk)

WWF

The World Wide Fund for Nature (WWF), founded in 1961, is the world's largest private nature conservation organization. It consists of 29 national organizations and associates, and works in more than 100 countries. The coordinating headquarters are in Gland, Switzerland. The WWF mission is to conserve biodiversity, to ensure that the use of renewable natural resources is sustainable and to promote actions to reduce pollution and wasteful consumption.

UNESCO

The United Nations Educational, Scientific and Cultural Organisation is the only UN agency with a mandate spanning the fields of science (including social sciences), education, culture and communication. UNESCO has over 40 years of experience in testing interdisciplinary approaches to solving environmental and developmental problems, in programs such as that on Man and the Biosphere (MAB). An international network of biosphere reserves provideS sites for conservation of biological diversity, long-term ecological research, and testing and demonstrating approaches to the sustainable use of natural resources.

The Royal Botanic Gardens, Kew

The Royal Botanic Gardens, Kew has 150 professional staff and associated researchers and works with partners in over 42 countries. Research focuses on taxonomy, preparation of floras, economic botany, plant biochemistry and many other specialized fields. The Royal Botanic Gardens has one of the largest herbaria in the world and an excellent botanic library.

International Society for Ethnobiology

The ISE is the largest professional organization dedicated to sustainable use of natural resources and community development, and has endorsed the compilation of this manual.

The People and Plants initiative is supported, financially, by the European Union, the Department for International Development (UK), the MacArthur Foundation and the National Lottery Charities Board (UK).

Department for International Development

NATIONAL LOTTERY CHARITIES BOARD

MACARTHUR FOUNDATION

DISCLAIMER

Contents

The 'People and Plants' Initiative

The 'People and Plants' Initiative aims to contribute to the sustainable and equitable use of plant resources by supporting ethnobotanists from developing countries who work with local communities to:

- study and record the use of plant resources, and return results for the benefit of communities
- resolve conflicts related to conservation and over-exploitation of plant resources
- promote sustainable methods of harvesting wild plants
- enhance the value of plant resources to local people.

Part of the support provided by People and Plants is in the form of production and distribution of publications. *Plants and Protected Areas* is part of a series aimed at providing practical methodologies on key conservation topics. Further types of People and Plants publication include Working and Discussion Papers, which provide information on specific case studies in ethnobotany, and issues of a Handbook containing contacts and succinct advice. Training videos on key methodologies in ethnobotany are also produced.

Field projects are undertaken for People and Plants by WWF and UNESCO, at present particularly in Africa (Kenya, Tanzania, Uganda and Zimbabwe), the Himalayas (including Nepal and Pakistan), south east Asia (focusing on Sabah, Malaysia) and the south Pacific (notably Fiji). These projects involve a mix of courses, participatory planning, awards, small grants and exchanges, producing integrated packages designed to meet local priorities in conservation and development.

People and Plants is a partnership between the World Wide Fund for Nature (WWF), UNESCO and the Royal Botanic Gardens, Kew.

Alan Hamilton
Plants Conservation Officer
WWF International

Panel of advisors

Pablo Eyzaguirre
International Plant Genetic Resources Institute
Via delle Sette Chiese 142
00145 Rome, Italy

Gary J. Martin
People and Plants Program
B.P. 262
Marrakesh-Medina, Morocco

Suzanne Nelson
Native Seeds/SEARCH
526 N. 4th Ave.
Tucson, AZ 85705
USA

Dianne Rocheleau
Department of Geography
Clark University
Worcester, MA 01610
USA

Julie Velasquez Runk
Smithsonian Tropical Research Institute
Apartado 2072
Balboa, Republic of Panama

Mark Wright
VSO Field Office
P.O. Box 137
Belmopan, Belize

Foreword

After earlier volumes on ethnobotany and plant invaders, this is the third in the series of People and Plants Conservation Manuals. It addresses the challenge of on-site conservation of useful plant resources in protected areas, as well as the maintenance of such resources in agricultural environments. As such, the manual examines concepts and methods derived from a wide range of scientific disciplines and resource management fields, including plant ecology, conservation biology, forestry, agronomy, ethnobiology and anthropology. But the *in situ* protection of plant resources is much more than a field application of scientific concepts. It is also a social process, that involves working closely with local communities. It entails recognizing the human dimension of protected areas and rural landscapes.

The sorts of philosophy and approach outlined in the manual have many points of shared concern with the biosphere reserve concept, promoted within UNESCO's Man and the Biosphere (MAB) Program. Biosphere reserves are much more than just protected areas, in that they seek to provide working examples for promoting both conservation and sustainable development. Such examples can only be credible if they express the social, cultural, spiritual and economic needs of society, and are also based on sound science. They can only work if they bring knowledge of the past to the needs of the future, and if they demonstrate in concrete terms how to overcome the problems of the sectoral nature of our institutions.

Field experience gained in putting concepts into action in several biosphere reserves in the US/Mexican borderlands figure prominently in the examples drawn upon by John Tuxill and Gary Nabhan in this manual. More generally they and other examples provide insights to *in situ* conservation that will prove useful to those in many parts of the world involved in managing plant resources in a broad regional context. Above all, the manual underlines the need for us to conserve both traditional knowledge and biological diversity, and to recognize the role and importance of them both in approaches to a more sustainable use of natural resources.

Pierre Lasserre
Director
UNESCO Division of Ecological Sciences, Paris

Preface

Over the past three decades, scientists, natural resource managers and environmental policy-makers have recognized that the conservation of biological diversity depends upon protecting and managing intact natural habitats. Such recognition has given greater importance and urgency to the international movement to establish and maintain biosphere reserves, national parks, forest reserves, wildlife sanctuaries and other protected areas. Lacking such refuges, biological diversity will continue to vanish from our planet in a steady stream of extinctions, a trickle in some places, a torrent in others.

Yet for one prominent component of biological diversity – the genetic variability contained within culturally valuable plant resources – conservation efforts have focused primarily on safeguarding these resources away from the habitats and cultural contexts in which they evolved. The off-site protection of plants in botanical gardens, seedbanks and other centralized institutions has met with many successes and prevented the imminent extinction of valuable plant gene pools. However, a comprehensive conservation strategy must complement off-site protection with on-site maintenance of plants in native wildlands and traditionally diverse agricultural landscapes.

Few management plans for protected areas currently address the conservation of native plants considered useful by human communities for their genetic potential, subsistence value, or ceremonial and social importance. This manual seeks to help to fill this gap by elaborating what we term an *in situ* (on-site) approach to useful plant conservation. The manual has two goals: first, to facilitate better management of protected natural habitats so that useful plant populations will not be lost or irreversibly degraded; and second, to illustrate promising approaches to conserving agricultural plant resources within the rural landscapes of their origin. The manual is intended primarily for conservation professionals, protected area managers, agricultural resource managers and others involved with managing plants, the habitats where they occur and other natural resources.

In Chapter 1 we describe the useful plant resources that all too often fall through the conservation safety net. Chapter 2 outlines an *in situ* approach to conserving useful plants in their native habitats, emphasizing the collaborative role of protected area managers, local residents and conservation professionals. Next, Chapter 3 describes approaches that can give voice to the concerns and knowledge of local communities, and that advance local participation in plant conservation efforts. In Chapter 4 we present an approach to setting priorities for conserving plants, and demonstrate how to plan for *in situ* management. Chapter 5 reviews approaches to monitoring and evaluating the status of plant resources that are the subject of management efforts. In the final chapter, we address the specific case of agricultural environments, and describe approaches that organizations and management agencies have developed for working with farmers and other local residents to conserve agricultural plant resources.

Designating an area as a park, refuge or biosphere reserve is only the first step towards the conservation of biological diversity. Sadly, some plant populations have continued to decline even in areas designated and maintained as protected, due to lack of detailed knowledge about the plants' ecology or the stresses they face. The guidelines this manual offers for *in situ* management are based on concepts and methods derived from plant ecology, forestry, agricultural sciences, conservation biology, ethnobiology, anthropology, rural development and natural resource management. The information required for managing plant resources in a local or regional context may come from a variety of sources, including local land managers or naturalists, traditional plant experts (such as herbalists) in neighboring communities, or consulting scientists.

Readers may notice an arid American tint to many of the examples presented in this manual, and we are particularly indebted to the Native American tribes of the US/Mexico Borderlands for teaching us much about plants and their conservation value. We have also been aided in the preparation of this manual by many organizations and institutions, including Native Seeds/SEARCH, the Arizona-Sonora Desert Museum, World Wildlife Fund-US, World Wide Fund for Nature, Sonoran Institute, Centro Ecologico de Sonora, Conservation International, USDA Forest Service, University of Arizona Office of Arid Land Studies, Duke University Center for Tropical Conservation, Center for Plant Conservation, University of Wisconsin Institute for Environmental Studies, Worldwatch Institute, Biodiversity Support Program, the Smithsonian Tropical Research Institute, the United Nations Educational, Scientific and Cultural Organization, and the UN Food and Agriculture Organization.

Among the individuals who have kindly reviewed drafts of this material and/or supplied information are Janis Alcorn, Barbara Dugelby, Pablo Eyzaguirre, Raymond Guries, Jack Kloppenburg, Aiah Lebbie, Crescencia Mauer, Brien Mielleur, Suzanne Nelson, Don Norman, Dianne Rocheleau, Humberto Suzan, Julie Velasquez Runk, Jorge Ventocilla, Andrew Vovides, Don Waller, Dan Whyner, Garrison Wilkes, Mark Wright and the *People and Plants* Conservation Manuals team: Bob Carling, Alan Hamilton, Gary Martin and Martin Walters. To them, and to other friends, family and colleagues who have helped this project come to fruition, we extend our heartfelt thanks.

1

Why conserve plant resources *in situ*?

Figure 1.1 A community-initiated botanical nursery for endangered cacti and other rare plants at Miquihuana, Tamaulipas, Mexico. Rare plants found on community land that is due to be developed for roads, agriculture or other activities are transplanted to this site, where the plants are monitored and tended by community members. (Photograph: G.P. Nabhan.)

Any native plant can be considered a resource for humankind – and for associated animal and plant species – in that it contributes to the health and stability of the ecological community in which it occurs. However, in any flora there are certain plants that are more intimately linked with human welfare. These can generally be divided into two groups: those which are used directly for home consumption or local elaboration of products, or sold to generate income; and those which are used indirectly as sources of genes, cells, cuttings, or novel chemical compounds for products developed by agricultural, pharmaceutical, or related industries.

For the purposes of this manual, plants which are used directly by local communities for food, forage, fiber, timber, medicine, ceremony, symbol, or income will be called **ethnobotanical** resources. Plants which are processed or manipulated to extract genes or DNA for crop improvement, cells for tissue culture, or chemicals for precursors of pharmaceuticals will be called **phytogenetic** resources. When we wish to talk about both categories in general, we will refer to them as **useful** plant resources, or simply 'plant resources'. Keep in mind that some plant resources may not fit neatly into either the ethnobotanical or phytogenetic category. Examples include commodity crops grown for international export or national consumption, or plants that are sold in a regional market by one community for direct use by another. In addition, some useful plants are phytogenetic resources to one person and ethnobotanical resources to another, a point to which we return in later chapters.

This manual will refer to both wild species and domesticated plants, such as crop **landraces** or **folk varieties**. We use the labels **wild** (for plants that grow and reproduce without human interference) and **domestic** (plants selected, sown, tended and harvested by people) loosely, since there is not always a clear distinction between the two. For instance, certain wild herbs of the genus *Amaranthus* prosper in farmers' fields worldwide, and are often tended and harvested as nutritious green vegetables. While these herbs are not planted by farmers, they may have been selected to suit the ecological conditions of traditional croplands and other human-influenced vegetation; in scientific terms, such plants are called **agrestals** (field associates) or **ruderals** (roadside associates).

All people are in some way dependent upon both ethnobotanical and phytogenetic resources for their continued survival. High-yielding varieties of rice that feed hundreds of millions of people around the world owe their disease resistance to genes from a single accession, or sample, of the wild rice *Oryza nivara* collected in central India [179]. More than one-quarter of modern pharmaceuticals contain plant chemicals as active ingredients, and millions of people worldwide still rely upon herbal remedies gathered from wild habitats and home gardens for their primary health care [173]. Plant conservation, to be sure, is not simply an esoteric subject of interest only to botanists – it should be of concern to all of us who eat, use or otherwise benefit from contact with plants.

As human populations have grown exponentially over the past two centuries, they have modified habitats, translocated species, and altered the distribution and abundance of plant resources on a scale unprecedented in history. The 248 000 scientifically identified plant species make up about one-sixth of all species presently named, but estimates suggest that up to 50 000 species of plants may become extinct during the next three decades unless there is significant action to reduce pressures upon them [249]. Many plant resources that formerly were widespread and abundant enough to harvest routinely will be relegated to a precarious existence in densities too low to sustain certain uses. They may become totally dependent upon wise human intervention for their continued survival.

1.1 *In situ* and *ex situ* approaches to plant conservation

Two kinds of human intervention have aided the conservation and continued existence of threatened plants. We refer to them as *ex situ* (removed from habitat) and *in situ* (in habitat) approaches (Boxes 1.1 and 1.2). In practice, between these two conservation strategies lies a gradient of approaches to effective conservation. For example, a rare plant can be propagated along an interpretive trail within its natural range, where it may still exchange genes with surviving wild populations. Although its seed may have come from an *ex situ* gene bank, its propagation has placed it in an *in situ* context very similar to that in which its ancestors evolved.

Box 1.1 Ex situ maintenance

The removal of plants from native habitats, for translocation off-site to intensively managed facilities, represents *ex situ* conservation (Table 1.1). This conservation strategy may use botanical gardens, arboreta, seed banks, clonal repositories, or more sophisticated (and expensive) greenhouses and laboratories for frozen storage of propagules or tissue culture of cells (Figure 1.2). It can also involve the informal culture of wild plants in nurseries, home gardens, or community botanical gardens (Figure 1.3). Of course for traditionally cultivated crops, such home or community gardens are *in situ* rather than *ex situ* conservation sites – illustrating the overlap that exists between these terms.

During colonial eras, *ex situ* propagation fostered the rapid spread of economic plants from one continent to another. Rubber and cacao were taken from Brazil and established in Southeast Asia and West Africa, sugarcane and bananas were tranferred from Southeast Asia to the Caribbean and Central America, and coffee moved around the world from East Africa to Latin America by way of Indonesia. From cuttings of plants nurtured in botanical gardens to hybrid seeds of cereals developed in agricultural experiment stations, *ex situ* germplasm collections have revolutionized agriculture and forestry over the last 500 years.

Table 1.1 Comparing *in situ* and *ex situ* conservation approaches

	Purpose	Examples	Benefits
In situ	To maintain species and populations in the ecological and cultural context in which they have evolved	National parks Sacred groves and religious sites Wildlife refuges Nature reserves Community woodlots, grazing lands	Species, populations and gene pools continue to evolve and adapt Ecological interactions between species and species assemblages (communities) maintained All native species protected, not just those with known value Ecological services and processes (e.g. pollination, water filtration) maintained
Ex situ	To maintain species populations and genes in intensively managed, easily accessible environments	National gene banks Commercial seed banks Botanical gardens Agroforestry collections Home gardens	Species/varieties threatened *in situ* can be cultivated and propagated for reintroduction Convenient access to germplasm for breeders, farmers, herbalists, researchers and other users Species and varieties are readily available for conservation, research and education

For several decades now *ex situ* conservation has been pursued internationally in a highly organized fashion, through government-supported seedbanks and institutions such as the Consultative Group on International Agricultural Research. Most of their attention has been directed at agricultural phytogenetic resources, but wild plants also are being maintained *ex situ* on an increasingly large scale. One major effort is the Millennium Seed Bank at the Royal Botanic Gardens, Kew, UK. This *ex situ* facility focuses on arid-adapted wild species from tropical and subtropical regions, and currently stores seeds of over 4000 species from more than 100 countries. Kew has a campaign underway to expand this collection to 25 000 species by the year 2010. When achieved, this total will represent approximately 10% of the world's known arid tropical and subtropical flora.

Figure 1.2 A field germplasm collection of cacao or chocolate (*Theobroma cacao*) trees near Port-of-Spain, Trinidad, managed by the University of the West Indies. Field gene banks are used for species like cacao whose seeds are recalcitrant, and cannot be stored for long periods under the dry, cold conditions of a conventional seedbank. (Photograph: R.P. Guries.)

Figure 1.3 Dooryard garden at Quitovac, a Sand Papago (Hia Ce'ed O'odham) community in Sonora, Mexico. Though such gardens are small in size, they often contain a diverse complement of local crop varieties grown for food, medicine, ornamental value and many other uses. (Photograph: G.P. Nabhan.)

This unprecedented assemblage of drylands species will not be merely held in storage, but will form a working *ex situ* collection, supporting a variety of education, training and research activities. Already, the Millennium Seed Bank distributes about 2000 small samples of seeds from its collection to researchers and organizations worldwide (the samples are sent free of charge to individuals in developing countries). Some seed recipients are studying the systematics of arid-lands plants, while others are conducting field trials to assess the potential of different tree species for arid-lands agroforestry. Certain species in the Millennium Seed Bank may also have commercial potential. For this reason, any use of seed samples for commercial purposes must be approved in advance by Kew, and in the event of financial success must adhere to the Kew policy of returning a portion of commercial profits back to the country in which the seed sample was collected. This question of how to recognize **intellectual property rights** to commercially valuable phytogenetic resources is a hotly debated aspect of both *ex situ* and *in situ* conservation, and is discussed further in section 6.11 of Chapter 6.

Box 1.2 In situ management

The protection of plants in their native habitats, termed *in situ* conservation (Table 1.1), is aimed at keeping intact the ecological relationships between species and, in some cases, cultural relationships between people and wild species [106, 107, 243]. This kind of strategy may involve formal protected areas recognized by governments, such as national parks, international biosphere reserves, state forest reserves, and private wildlife sanctuaries or conservation areas. It also includes informally recognized or culturally protected areas, such as sacred groves, springs and mountains, and reserves for medicinal plants where communities have decided that no other extractive activities are allowed. *In situ* protection may take the form of traditionally tended hedgerows and woodlands where clearing, building or grazing are restricted. For locally adapted crop varieties, *in situ* conservation can be the simple act of selecting the most favorable seeds from one harvest for replanting the following season, year after year.

In most parts of the world, *in situ* conservation for useful plants has tended to be locally and informally organized. Efforts by government institutions or international organizations to promote *in situ* plant conservation are relatively recent, but expanding. The International Plant Genetic Resources Institute, which has long been at the center of *ex situ* conservation, began a five-year project in 1995 to identify and promote scientific guidelines for managing phytogenetic resources *in situ*, and to strengthen the capacity of government agencies to implement *in situ* conservation. In some cases, plant users are also organizing locally to support *in situ* plant conservation through formal channels. For instance, members of the Belize Association of Traditional Healers have long used wild plants in treating patients, but through the Association they are now working to establish and manage a forest reserve in Belize specifically for medicinal plants [17].

Not until early in the 20th century did plant explorers such as N.I. Vavilov, Harry Harlan and Efrain Hernandez Xolocotzi recognize that some of the same plants they were collecting for genebanks were also becoming extinct in their native settings [79, 234]. Although the provision of raw materials for crop improvement has remained the primary goal of gene banks, another goal has more recently become apparent: genetic conservation itself [111, 149]. Such conservation may allow future reintroductions to habitats where a plant has been lost, or provide sources for seeds of folk varieties that farmers previously abandoned but now seek to grow again.

Some plants now extinct in the wild survive only because of their timely collection and careful stewardship in botanical gardens and seedbanks (Figure 1.4). However, many germplasm collections lack sufficient genetic variation to be truly representative of prevailing gene frequencies in a species' original range. Indeed, these *ex situ* stored vestiges may be so impoverished that some scientists claim they are no longer functional species [137]. In some cases, all that remains are a few vegetative cuttings, or **ramets**, of a single genetic individual, or **genet**. Some scientists doubt that such vestiges of a wild species would necessarily survive in the wild were their habitats to be restored and their progeny reintroduced. To date, rare plant reintroductions and translocations have had limited or at best mixed success [70], although better horticultural techniques and improved ecological planning can increase their chances [74]. However, reintroductions are inevitably far more costly than habitat protection would have been decades earlier.

A new generation of conservation biologists is now taking the genetic conservation objective of gene banks as seriously as that of serving crop improvement programs. Scientifically based guidelines are being developed for the adequate and efficient capture of the existing genetic variation in declining species [44]. These allow samples to be propagated and reintroduced in ways that minimize subsequent reductions in genetic diversity [74, 75].

Sampling strategies that are well grounded in population genetics improve the probability of capturing a high percentage of a species' remaining alleles. However, it is virtually impossible to predict whether coadapted gene complexes will remain together under the artificial conditions of greenhouse breeding, seed

Figure 1.4 *Franklinia alatamaha*, a species known from only one location in the southeastern United States, where it was collected during the early 19th century; it has never since been relocated in the wild. The tree now exists only in botanical gardens, arboreta and similar facilities. (Photograph: R.P. Guries.)

bank storage and botanical garden exhibit. Despite attempts to the contrary, genetic landscapes can never be held perfectly stable when stored off-site [110]. Propagating plants *ex situ* may result in artificial selection; for instance, sowing seeds of a dunes-adapted plant population for regeneration on the clay loam soil of an experimental farm may select against adaptations to sand. Gene frequencies can change as a result of differential losses in storage, during germination and establishment, or during maturation and harvesting in experimental fields [196].

These genetic limitations of *ex situ* collections are all the more sobering given that many seed banks and germplasm repositories are underfunded and have limited resources for documentation, maintenance, and upkeep of collections. Many samples are not grown out frequently enough to maintain their viability, and inadequate labelling of vouchers is an ongoing problem. Under present constraints, the additional expertise and costs required to implement genetic guidelines for germplasm sampling and maintenance are likely to be met by only a subset of *ex situ* facilities [98].

1.2 Integrating *in situ* and *ex situ* strategies

As the limitations of *ex situ* genetic conservation have become more apparent, there has been renewed attention given to *in situ* conservation strategies [160,

38]. In practice, these two strategies can be complementary. For example, when the Mexican psychotropic plant *Salvia divinorum* became rare in the wild, Mazatec herbalists learned to propagate it in secluded settings within its former range, thereby increasing its numbers for ceremonial use [207]. More recently, botanists have taken frozen stored pollen from propagated clones of a rare century plant, *Agave arizonica*, and used it to pollinate the few isolated individuals remaining in the wild (Figure 1.5). Without this *ex situ* pollen source, the flowers of these *in situ* survivors would never have set fertile seeds [105].

A more balanced view is emerging that recognizes the comparative advantages and limitations of *in situ* and *ex situ* conservation strategies [74, 160]. There is growing support for employing both strategies together as an integrated approach that draws upon each strategy's particular strengths. Such an approach will offer the best possible response to the specific threats facing plant resources in any one area or location. It is no longer sufficient to assume that a plant will survive well into the future just because it grows within a designated park or reserve. Neither formal scientific nor indigenous conservation strategies have dealt sufficiently with global climate change, habitat fragmentation, fast-spreading exotic species and the many other challenges that plants will face in the rapidly changing environment of our planet [186]. Instead, the **adaptive management** process outlined in this manual encourages ongoing observation,

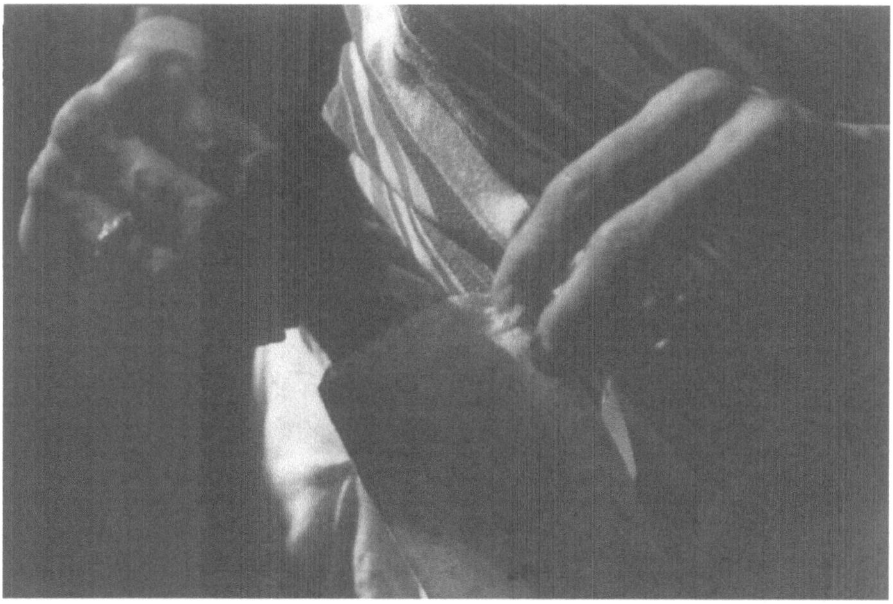

Figure 1.5 A pollen sample taken from *Agave murpheyi* for frozen storage *ex situ*. This species, a close relative of A. *arizonica*, is known from only a handful of locations in the Sonoran Desert and is genetically depauperate. (Photograph: G.P. Nabhan.)

monitoring and evaluation to allow for changes in management as new threats or conditions become apparent.

In some cases, adaptive plant management can remain relatively passive, with botanists or resource managers working with local residents and landowners to monitor the effects of natural landscape processes on the regeneration of plant populations of concern. In other cases, more active intervention may be necessary, including controlled burning, *ex situ* propagation, population enhancement, translocation, pollinator reintroduction, or removal of competitors. Each action must be assessed with regard to its actual benefits to the species of concern and to the biotic community as a whole, relative to its cost. *In situ* conservation, then, is far more complex than the simple designation of a protected area where rare plants occur. It must trigger a series of constantly shifting interactions between plants, people and other animals that will shape and re-shape the destiny of a threatened plant resource for years to come.

2

In situ plant conservation: who is involved?

Figure 2.1 Researchers and local residents conducting an inventory of trees in a sacred forest grove in Sierra Leone. Sacred groves are an important local conservation tradition throughout many regions of the world. (Photograph: A.R. Lebbie.)

We began this manual by identifying the intended audience, and will now consider at greater length a related question. When it comes to routine, on-the-ground activities, who are the people involved in conserving useful plants in their native habitat? Successful *in situ* plant conservation will be a collaborative process between three groups:

- professionals or researchers who focus on plant conservation, including botanists, agricultural scientists, ethnobiologists, and social scientists with particular interests in plant resource use;
- local residents, particularly harvesters and land stewards, or, for agricultural resources, farmer-custodians;
- natural resource managers, including protected area administrators, technical staff (both biological technicians and social science technicians), ranger-naturalists and community extension workers.

Under the first category fall scientists who study plants and their uses, representatives from conservation funding agencies, advocates from non-governmental organizations, government scientists responsible for conservation policies, botanical garden researchers and many other individuals. They are usually trained in the natural or social sciences (or both) and are in some way concerned about the loss of plant resources, but they tend not to be directly involved in day-to-day land management. Many such professional researchers and advocates (including ourselves) have a deep and abiding personal commitment to protecting plants and their habitats. We have chosen to make this a focus of our life's work. Individuals in this category form a collective – if often informal – international 'support team' that can provide advice and expertise, particularly on technical matters, to local residents and natural resource managers alike.

People in the latter two categories have a central role in plant conservation because they are engaged daily in managing valued plant resources and the habitats they occupy. For centuries, local residents have been the primary managers, manipulators and custodians of useful plants. In most cultures, there are exceptional plant specialists who know details of the life cycles and ecology of hundreds of species [26]. Protected area and natural resource managers – more recent arrivals on the scene – are charged by national and state governments with managing lands to conserve various natural features, including plant populations. The following discussion explores why we cannot expect to undertake *in situ* conservation without fully engaging both local residents and professional resource managers. In addition, we examine briefly the complex historical relationship between the two groups.

2.1 Why focus on local people?

We are convinced that any effort to conserve plant resources *in situ* must have the full participation and support of local residents. There are several well founded reasons for giving extensive consideration to local concerns, outlined in the following sections.

2.1.1 *Global phytogenetic resources are also local resources*

Many plants valuable for developing new crop lines, pharmaceuticals and other modern agroindustrial products are also plants that have sustained local communities for centuries. In some cases, the linkage is obvious; for instance, traditional crop varieties that form the bulk of the raw material used by modern plant breeding programs have for centuries been cultivated and developed directly by rural residents. Conservation efforts for such phytogenetic resources must direct paramount attention to the custodians who have shaped and maintained the plants and their agricultural contexts for generations.

Other useful wild plants may receive no more manipulation or attention than occasional harvesting, yet still be of great import to local residents as medicines for curing ailments, as essential components of religious ceremonies, or as valued foodstuffs (Box 2.1). A conservation effort that does not acknowledge the local importance of such resources risks becoming a point of contention for residents rather than a program worthy of their support. In cases where residents are denied access to long-utilized resources, such efforts can lead to severe hardship and discontent.

2.1.2 In situ *conservation can benefit from local residents' knowledge*

Many surveys have recorded high levels of biological diversity – and local residents'

Box 2.1 Examples of local reliance on wild plants

- Across the deserts and savannas of Africa, people harvest over 60 species of wild grasses for food [100]. The Zaghawa, pastoralists from Chad and the Sudan, regularly harvest up to seven different annual wild grasses, including shama millet (*Echinochloa colona*), wild rice (*Oryza breviguIata*), and wild tef (*Eragrostis pilosa*), all close relatives of cultivated grains. Wild grains are eaten directly and also used to brew beer [230].
- The Kuna, an indigenous group from Panama, employ a wide variety of wild plants in their traditional medicine. In ethnobotanical interviews, two Kuna healers indicated that between them, they use over 70 species of wild plants in curing ceremonies [96].
- In rural northeast Thailand, wild foods found in forests, rice paddies and field margins typically make up half of all food eaten during the wet season. These wild foods include fruits, vegetables, mushrooms, fish, snails and frogs [216].

detailed knowledge of this diversity – within landscapes traditionally managed and exploited by communities [41, 53]. However this local knowledge base has often gone unacknowledged by professional natural resource managers or administrative agencies [231]. Raymond Dasmann [61] reminds all of us to respect people who have a long history of use or occupancy of an area, for they

> have a familiarity with its species, communities, and ecological processes which cannot be readily gained through surveys, inventories, or baseline studies by experts from elsewhere. In particular, long term trends or fluctuations in abundance and distribution of wild species, past influences and changes, values and usefulness for human purposes can be determined most easily from local people. Consultation with these people is essential to gain the knowledge important for both conservation and for the avoidance of conflict.

The traditional knowledge of local biodiversity referred to by Dasmann has been documented for a wide range of rural cultures through ethnobiological studies. Tzeltal Maya campesinos in tropical Mexico recognize over 1200 species of plants and animals [108, 27]. Even in a desert where species diversity is relatively low, the River Pima in Arizona have names for over 750 plants and animals [187]. Several ethnobotanical surveys of rural communities in the Amazon Basin have found that residents can identify uses for more than 90% of the tree species found near their communities, a number all the more impressive when one realizes that certain Amazonian forests contain over 300 tree species per hectare [169, 172, 178].

Other studies have disclosed extensive indigenous knowledge of soil properties, ecological communities and vegetation successional dynamics [91, 153]. Managers may also find that local people know a great deal about the distributions and causes of decline of rare, highly endemic, or economically important plants not yet well studied by western scientists (Box 2.2). For instance, we estimate that Seri, O'odham and Yuman indigenous communities know about one-third of the rarest Sonoran Desert plants better than do academically trained field scientists. When managers respect and incorporate local people's knowledge and experience, opportunities for collaborative *in situ* management expand enormously.

There is no guarantee that such opportunities will always be there in the future. Researchers, community members and resource managers alike have observed that traditional knowledge of plants, animals and other elements of nature can disappear even more rapidly than the organisms themselves [152]. In such situations the ecological understanding held by older generations is not passed on to younger community members, who forsake the stories and training of their elders for the trappings of industrial society. This trend has been documented for knowledge of medicinal plant use [173], agricultural practices [236], and traditional resource management [47, 163]. Even people who historically have demonstrated a remarkable determination to resist outside pressures for

Box 2.2 Local ethnobotanical collectors at Mount Kinabalu

In Malaysia, a collaborative ethnobotanical program called *Projek Etnobotani Kinabalu* (PEK) has trained local botanical collectors in several ethnic Dusun communities surrounding Mount Kinabalu National Park. For over 140 years, the flora of this park has also been extensively collected by professional botanists. Researchers recently compared the relative performance of professional and community botanical collectors in documenting species of palms (family Arecaceae) in the Mount Kinabalu region [134].

Review of a database of European and Asian herbaria records revealed that between 1856 and 1992, over 200 botanists visited Kinabalu and collected 47 species of palms, representing 10 genera. Between 1992 and 1996, 12 local Dusun collectors documented 68 palm species, representing 19 genera. Out of a total of 75 palm species known from Kinabalu, seven have been collected only by professional botanists, while some 28 species are documented solely in the plant presses of Dusun collectors. Put another way, after only 3½ years, local collectors had surveyed 90.7% of the known palm diversity, while after 136 years, visiting botanists managed to document only 62.7%.

PEK advisers attribute the stellar performance of Dusun ethnobotanical collectors to several factors, most prominently the greater geographic coverage attained. Local collectors documented palms at sites all around the Kinabalu massif, while visiting collectors tended to work in the southern section of the park, along established trails and at higher elevations. Furthermore, local collectors targeted not just wild species, but cultivated and informally managed species as well. Local collectors also brought to their work a life-long familiarity with organisms, localities and microenvironments around Mount Kinabalu. The experience of PEK suggests that the most accurate and complete documentation of a region's flora and fauna will be achieved with biological surveys implemented jointly by local and professional collectors. Such collaboration will make the most of the respective strengths and knowledge of each set of collectors [134].

cultural change may now face a growing gap in ecological knowledge and practices between older and younger generations [20, 46]. Social scientists and other researchers still do not understand well the reasons why, when and how traditional knowledge and practices persist in a culture or community [146]. Nevertheless, when conservationists are cognizant of cultural changes ongoing in local communities, they can develop management approaches more likely to strengthen traditional ecological knowledge as well as conserve valued plant populations.

2.1.3 *Local conservation practices offer a foundation for* in situ *conservation*

Local resource management traditions can be a productive starting point for collaboration between land managers, conservation professionals and local residents. Ethnobiologists have identified numerous cases where communities have developed sophisticated practices to safeguard areas of undisturbed habitat, remnant populations of useful plants, and other valued natural resources [47, 123, 171]. Such practices represent locally initiated resource conservation policies, and may be more deeply integrated into the cultural fabric of local society than are the conventional land management policies of nation-state governments. For example, many cultures throughout Africa, Asia and the Pacific have traditions of designating sacred groves – areas of forest where villagers are strictly prohibited from clearing trees or otherwise disturbing the vegetation (Box 2.3).

Box 2.3 African sacred groves as plant conservation sites

Throughout sub-Saharan Africa, rural villages maintain parcels of native forest as sacred groves, guarded and overseen by specialized social groups within village communities. Along the coast of Kenya, sacred groves called *kayas* are overseen by councils of elders. In Sierra Leone, the groves are maintained by villagers skilled in the use of particular medicinal plants, who are usually organized along gender lines into secret societies – groups whose membership is restricted to properly initiated individuals.

Wherever they occur, sacred groves play a powerful role in traditional village life. According to biologist Aiah Lebbie and forester Raymond Guries [123], Sierra Leonean sacred groves are sites for initiation ceremonies, spiritual rites, adolescent education, training grounds for herbalists, and sources of medicinal plants; some even serve as traditional hospitals where sick individuals are taken for specialized treatment.

While most groves are small – rarely more than 10 ha, and often only one or two hectares – they possess a disproportionately large conservation value. Because of widespread deforestation and agricultural expansion, in many areas sacred groves now protect the best remaining fragments of mature forest (Figure 2.1). Access to sacred groves is usually highly restricted; typically the explicit permission of village elders or society leaders is required to enter the groves. Plants and animals in the groves are protected by strict taboos, which strongly discourage overexploitation. For some African flora and fauna, sacred groves may be their last refuge. In interviews that Aiah Lebbie conducted with village herbalists, several assistants reported that certain medicinal plants formerly abundant in the once-extensive forest could now be collected only in sacred groves.

The long-term future of sacred groves in village life is far from assured. Since 1986, botanists Ann Robertson and Quentin Luke have carried out surveys of *kayas* within the coastal forests of Kenya [192]. They report that these sacred groves face widespread degradation and even outright destruction from tourism development, mining, and other activities. In some cases, younger villagers no longer attach the same significance to sacred groves that their elders do. The small size and isolation of most sacred groves increases the likelihood that they will lose biological diversity as species now restricted to them suffer local population extinctions.

Despite these hurdles, sacred groves continue to hold potential for *in situ* conservation collaboration between local residents and conservation professionals. In Kenya, the government has declared a number of *kayas* as national monuments, and the National Museums of Kenya has established a Coastal Forest Conservation Unit to promote their conservation [248]. Sacred groves are already widely accepted by many local residents as an important part of village life. Conservation professionals can offer support to sacred grove custodians who safeguard and maintain local populations of medicinal and other useful plants. In so doing, they may bolster a traditional conservation approach as well as help to maintain declining plant populations.

Of course, not all communities have equally intricate or sophisticated conservation traditions. The land management techniques practiced by a village of recent immigrants unfamiliar with the local biota may be far less effective in maintaining the native flora and fauna than are the practices of a long-established community occupying traditional lands [157]. However, by building wherever possible on existing local resource management practices, formal *in situ* conservation programs gain several advantages. When conservation guidelines are developed and implemented in concert with long-standing local taboos, they are more likely to be acknowledged and supported by residents. In addition, like local knowledge systems, many traditional conservation practices are currently eroding with alarming rapidity as rural cultures face a barrage of social change. By adding social prominence and enhanced status to traditional practices, formal conservation efforts can help to influence whether or not such practices are maintained and adhered to by younger generations in a community.

2.2 Why emphasize protected areas?

No matter where they are working, *in situ* conservation advocates must coordinate their efforts with national and regional governments as well as local residents. In most nations, governments sponsor conservation efforts by creating natural resource management agencies and establishing protected area systems. Even when established local conservation practices like managing sacred groves are in place, there are several reasons why conservation of plants and their habitats can be made more successful with the careful involvement of protected area managers and resource managers.

To begin with, governments and other regional institutions tend to be better positioned to resolve threats to plant resources that originate outside of local communities. When communities enter into regional cash economies and undergo rapid cultural change, local residents often encounter outside pressure to exploit their lands for short-term gain. Under such conditions, it becomes increasingly difficult for local conservation practices alone to guard against habitat degradation and overexploitation of plant resources. For instance, rural communities from New Guinea to Brazil have been persuaded by logging operations to sell trees from their forests, for which loggers typically offer only a fraction of the timber's fair market value and a degraded forest in return [79, 86]. When implemented effectively, national protected area regulations can provide greater legal and political weight to counterbalance outsider interests. Unscrupulous governments, of course, often facilitate rather than block the predatory actions of well connected outsiders who covet the lands or resources of rural communities [164, 217]. In either case, such weight – whether positive or negative – is difficult for local conservation bodies to bring to bear on their own.

A role for protected area managers is further warranted because certain ecological relationships important for maintaining useful plant populations – and ecological communities as a whole – unfold across regional or even continent-wide scales. For instance, some plant resources are pollinated or dispersed by animals that migrate over long distances. In northern Mexico and the southwestern United States, populations of nectar-feeding bats vital for pollinating many prominent plants are in decline, and their rescue may depend as much on regional as on local conservation action (Box 2.4).

Box 2.4 Long-distance pollinators need regional protection

As they migrate between their winter roosts in central Mexico and summer breeding grounds along the US/Mexico border, lesser long-nosed bats follow a 3000-mile-long 'nectar corridor' provided by the sequential flowering of at least 16 plant species. These plants include several species of agaves harvested for the tequila industry, tree morning glories, and arborescent cacti whose fruits are consumed by rural Mexican and US residents alike. Near each site, it is possible to estimate how many flowers or plants a roosting population may need during its stay, and to manage vegetation to sustain these numbers [168]. However, overall conservation of the bats is more complicated than that.

Over time the bats' nectar corridor has become increasingly fragmented [42]. On US rangelands, ranchers attempting to control shrub growth apply powerful herbicides that can kill nectar source plants. Further south in Mexico, ranchers have converted vast stretches of diverse thornscrub into pastures planted to a single exotic grass. In addition, bats have seen their supply of agaves dwindle in parts of the Sierra Madre foothills due to overharvesting by bootleg mescal producers. Bats may also face direct poisoning threats from DDT and endocrine-disrupting pesticides applied to Mexican farms growing cotton and vegetables for export to the United States. Finally, ranchers have begun dynamiting the bats' winter roosts to kill vampire bats, which occasionally feed on the blood of their cattle.

As the long-nosed bats travel their migratory routes, they efficiently pollinate plants on a scale that would be impossible to duplicate through any other means. Protecting the bat migration as an ecological process has become the goal of collaboration between a number of Mexican and US natural resource agencies and conservation groups. While such conservation is local at any one site, it must also be regionally coordinated across the bats' entire range, amongst all people who live alongside and benefit from their presence. The international migratory bat commission recently formed in North America is mapping and monitoring more than 1500 roosts along three species' corridors.

When large-scale ecological interactions are involved, local action to protect a plant resource in one region may be undermined by ill-conceived land use practices tens or hundreds of miles away. Regional protected area systems and government resource management agencies have the potential to lead the way in coordinating conservation efforts over a wide area. Their partners in this task may include non-profit organizations, associations that link resource user groups in different areas (such as associations of traditional healers in African countries), and researchers who monitor species across their entire range. All such institutions and individuals are important partners in the conservation of ecological phenomena such as migratory wildlife, seasonal flood regimes, groundwater recharge patterns, and species' distributional shifts due to changing climate and rainfall.

2.3 Protected area approaches to plant conservation

While both local residents and natural resource managers have much to contribute to plant resource conservation, their interests and perspectives are not wholly equivalent to those of conservation professionals. Plant conservationists have their own priorities, particularly for sites that are centers of plant diversity or endemism [251] or unusually rich in globally rare species or useful plants. In some cases, conservationists have generated broad support to have sites designated as protected areas specifically to safeguard plant resources. As early as the 1950s, the former Soviet Union established phytogenetic reserves in the

Caucasus mountains to protect stands of wild apples, peaches and pistachios [107]. More recently, the rediscovery of two endemic perennial relatives of corn, *Zea perennis* and *Zea diploperennis*, in the Sierra de Manantlan mountains of Jalisco, Mexico, spurred the declaration of the area as an international biosphere reserve [109]. In India, the government has designated a forest reserve to safeguard several wild species of *Citrus* [213].

Yet for the most part, *in situ* plant conservation will continue to depend upon protected areas established for other reasons, such as the presence of rare animals, scenic beauty, protection of archeological sites or timber reserves, general biodiversity value, unique geologic features, or watershed maintenance. As Table 2.1 illustrates, protected areas come in a range of different designations, each with a

Table 2.1 Official categories of protected areas, as determined by the World Conservation Union (IUCN) (adapted from McNeely and Miller, 1984)

Category	Management objectives
Strict Nature Reserve/ Scientific Reserve	To maintain natural processes and habitats in an undisturbed site for scientific study, environmental monitoring, education
National Park	To protect outstanding undisturbed natural and scenic areas for scientific, educational and recreational use. No extractive resource uses
National Monument/ Natural Landmark	To protect significant natural sites with unique characteristics. Relatively small areas focused on preserving specific features
Managed Nature Reserve/ Wildlife Sanctuary	To protect significant species, biotic communities, or environmental features that require human management to persist
Protected Landscapes and Seascapes	To maintain significant cultural/natural landscapes characterized by traditional land use patterns and high opportunities for public enjoyment through recreation and tourism
Resource Reserve	A temporary designation to protect natural resources for future use and contain potentially damaging development until study and planning assessments can be completed
Indigenous Reserve/ Cultural Landscape Area	To allow indigenous communities to continue traditional lifestyles while minimizing external pressures or undesired development
Multiple Use Management Area/Managed Resource Area	To provide for sustained production of water, timber, wildlife, pasture, tourism and other resources
Biosphere Reserve (international designation)	To conserve the diversity and integrity of natural ecosystems and safeguard the genetic diversity of species present. Human land uses integrated with natural areas
World Heritage Site (international designation)	To protect natural features of international significance. These sites are nominated by national governments

distinct management approach. *In situ* plant conservation – as well as many other management goals – will be most effective when done via administrative arrangements that acknowledge long-standing patterns of resource use by local communities. Management must understand and accommodate such traditional patterns where possible, rather than automatically barring all human activities.

We state this conviction firmly for several reasons. The previous pages have already outlined why local residents' involvement in plant conservation is valuable and desirable in many situations; indeed, it is very often essential. Such involvement on the part of residents is not likely to be forthcoming unless protected area managers are willing to acknowledge the past and present extent of their knowledge and resource use.

In addition, a large slate of useful plant resources – both phytogenetic and ethnobotanical – are linked through their evolutionary history with human cultural practices. Certain useful plants favor habitats produced by human disturbance, such as abandoned agricultural plots. For instance, in the eastern Brazilian state of Bahia, Afro-Brazilian folk healers rely heavily on secondary forests and other disturbed areas for their supplies of medicinal plants [237]. Near Lake Patzcuaro, Mexico, Victor Toledo, Cristina Mapes and colleagues [227] found that fully half of the plant, animal and mushroom species sold in local markets came from human-modified habitats rather than natural habitats (Table 2.2). Of course, there are also many species that can survive only in mature forest, unplowed savanna, or other minimally disturbed habitats. Moreover, as rural communities undergo social and economic changes, they may not always maintain traditional land management practices or cultural attachments to native plants and animals. The fact remains, however, that protected area networks which automatically exclude all traditional land uses run the risk

Table 2.2 Number of species (plants, mushrooms and animals) in indigenous markets near Lake Patzcuaro, Mexico, classified by use and ecosystem type (reprinted from Toledo, 1991, by kind permission of M. Oldfield and J. Alcorn)

Group	Natural ecosystems				Modified ecosystems			
	Lake	Forests	Shrublands	Grasslands	Cultivated fields	Fallow fields	Kitchen gardens	Total
Foods	19	9	6	5	15	9	26	89
Medicines	1	13	3	0	0	1	13	31
Ornamentals	0	2	0	0	1	0	2	5
Forages	0	0	0	0	2	0	0	2
Fuelwoods	0	4	0	0	0	0	0	4
Domestics	2	1	0	0	0	0	0	3
Work/Tools	0	4	0	0	0	0	0	4
Total	22	33	9	5	18	10	41	138
(Percent of total species)	(16%)	(24%)	(6%)	(4%)	(13%)	(7%)	(30%)	

of disrupting beneficial ecological and cultural relationships between people and biological resources. Without such relationships in place, native plants will vanish from the landscape all the more rapidly.

Of the protected area management approaches listed in Table 2.1, the biosphere reserve model goes the furthest towards reconciling habitat and species protection, and local residents' land use needs and cultural traditions. To understand why biosphere reserves are not yet the norm for conserving natural areas, the following section briefly reviews the history of protected areas establishment worldwide and examines the philosophical underpinning of biosphere reserves. Our review emphasizes the impacts felt by local residents – in many cases representing indigenous or ethnic minority cultures – as protected areas have been created on lands that residents long considered their own.

2.4 Protected area–local resident interactions: a historical synopsis

No one culture or society can claim to have a monopoly on the concept of conservation. Among nation-states, however, western conservation philosophies have supplied the dominant models for developing formal protected areas and parks policy [101]. These philosophies were initially based on the premise of an inherent separation between human society and nature; thus by definition, a 'protected area' had to preclude any active human presence [156, 170]. Western countries first began to put wildlands conservation into practice on a large scale during the late 19th century, when the United States created Yellowstone and Yosemite National Parks, the world's first national parks. (It is worth pointing out that, over time, protected areas in western nations – particularly in Europe – have expanded conceptually to include designations for landscapes actively inhabited and shaped by human communities, such as the national parks system of Great Britain.)

What may be less well known is that establishment of these two landmark parks also entailed the expulsion of Native American peoples. Miwok communities were resident in Yosemite, while bands of Shoshone visited Yellowstone on a seasonal basis [11, 52]. Also during this time, colonial administrations throughout much of Africa and Asia began to decree protected areas such as forest reserves and hunting parks. In virtually all cases these protected areas were similarly superimposed over territory that was traditionally farmed, fished, hunted upon, gathered over, and otherwise called home by indigenous cultures [170].

In their quest to preserve natural landscapes, conservation professionals failed to recognize that many habitats they perceived as natural and pristine actually resulted from traditional habitat modification activities, such as regular burning, pruning, or shifting cultivation [11, 57]. In the majority of cases where national parks and other reserves were decreed during the early and mid-20th century, cultural activities that modified habitats from their unspoiled state were summarily banned within park boundaries. In the eyes of many conservationists, these activities were, implicitly or explicitly, the very disturbances which caused wild plants and animals to require protection. Local residents who used these plants and animals as subsistence resources were forced to abandon their traditional activities in protected areas, and in many cases were physically evicted from reserves [126]. In some instances, to protest at such treatment, communities purposefully exploit or overharvest plant and animal resources they previously managed with care. Such cases resulted in polarized relations between local residents and protected area managers, each with different cultural assumptions about the natural world [115].

As a result of this historical experience and their own cultural traditions, many rural residents understandably do not agree with the common conservationist assumption that for a place to be wild, humans must be absent. The history of conservation efforts in India is one example that illustrates this clash between perspectives (Box 2.5).

Box 2.5 Conservation history in India

The history of conservation in India illustrates the experience of many rural communities with colonial conservation policies. British administrators in India during the late 19th century refused to recognize the validity of community-managed common lands. Although they recognized cultivated land as private property, British rulers insisted on designating common lands held in forest, fallow and pasture as state property, for which only the state had the right to manage and assign uses [83]. By the turn of the century, many communities were being denied not only their traditional right to manage common lands, but also rights of access to forest reserves and other common areas where residents had traditionally hunted wildlife and gathered plant products [52].

An unwillingness to recognize local residents as valid participants in wildlands management continued to permeate Indian resource management agencies well after independence in 1948. For example, the demarcation of Sariska National Park in 1984 encompassed the traditional territory of 24 villages, and instantly made them, by law, illegal residents [63]. Initially the Indian government attempted to resettle the villagers, but without success. It is only during the past decade that protected area managers at Sariska and other Indian parks have begun to concede the essential role of local communities in protected areas resource management. At Sariska, roughly half of the villages have begun to re-establish their traditional land management patterns, under the encouragement of a local NGO and with concurrence from park managers [212]. According to the NGO, this reversion to traditional management has improved villagers' livelihoods and increased the security of the natural forest.

In sum, many strictly protected areas have resulted in profound social disruption among local residents, who were either denied traditional land and resource use rights or forced to relocate [170]. As evidence of this disruption is increasingly acknowledged and recognized, growing numbers of conservationists and natural resource managers are reconsidering past approaches to wildlands protection.

2.5 An emerging protected areas philosophy: adoption of biosphere reserve principles

By the late 1960s, a growing call could be heard from researchers and indigenous peoples' advocates to entertain new approaches to protected areas management. Conservation organizations increasingly sought arrangements that took into account local residents' rights. In part this was based on recognition that, in many cases, local people had been effective stewards of biodiversity and attendant natural landscapes.

The term **biosphere reserve** first arose to describe a protected area (Fig 2.2) composed of a strictly protected core reserve bordered by **management** or **buffer zones** and **transition areas** allowing traditional land and resource use [19]. Buffer zone activities are limited to uses compatible with the protected core area: scientific research, training, education, ecotourism, and recreation. Transition areas (which are erroneously labeled as 'buffer zones' in some cases) aim to promote sound resource management by local inhabitants. Transition areas also provide a context in which to address the development needs of local communities.

Actual implementation of this model began in 1979 with the establishment of the first biosphere reserves within the 'Man and the Biosphere Program' of the United Nations Educational, Scientific and Culture Organization (UNESCO). Today there are over 360 formally designated biosphere reserves in over 85 nations worldwide [19, 240]. In addition, biosphere reserve concepts such as buffer zones have been applied to many other protected areas [214], and the concepts are endorsed by a broad spectrum of nature conservation organizations. Even the Nature Conservancy, a US-based group focused on biodiversity preservation, now allows much more community involvement in its 'bioreserves' than it did in its first 40 years of establishing 'conservancy areas.'

With their integrated design, biosphere reserves offer a compelling conceptual approach to protected areas establishment. Under this approach, land managers can support local residents' land use activities when they are diversity-enhancing, and press for changes in use patterns that are detrimental to natural habitats and native species. We have already noted that the distributions of a wide variety of plant resources are closely linked to habitat mosaics produced and influenced by centuries of human handiwork. Biosphere reserves honor patterns of cultural intervention, such as hillside terracing and raised fields in wetlands. Therefore, they can serve to maintain wild plant resources that might otherwise

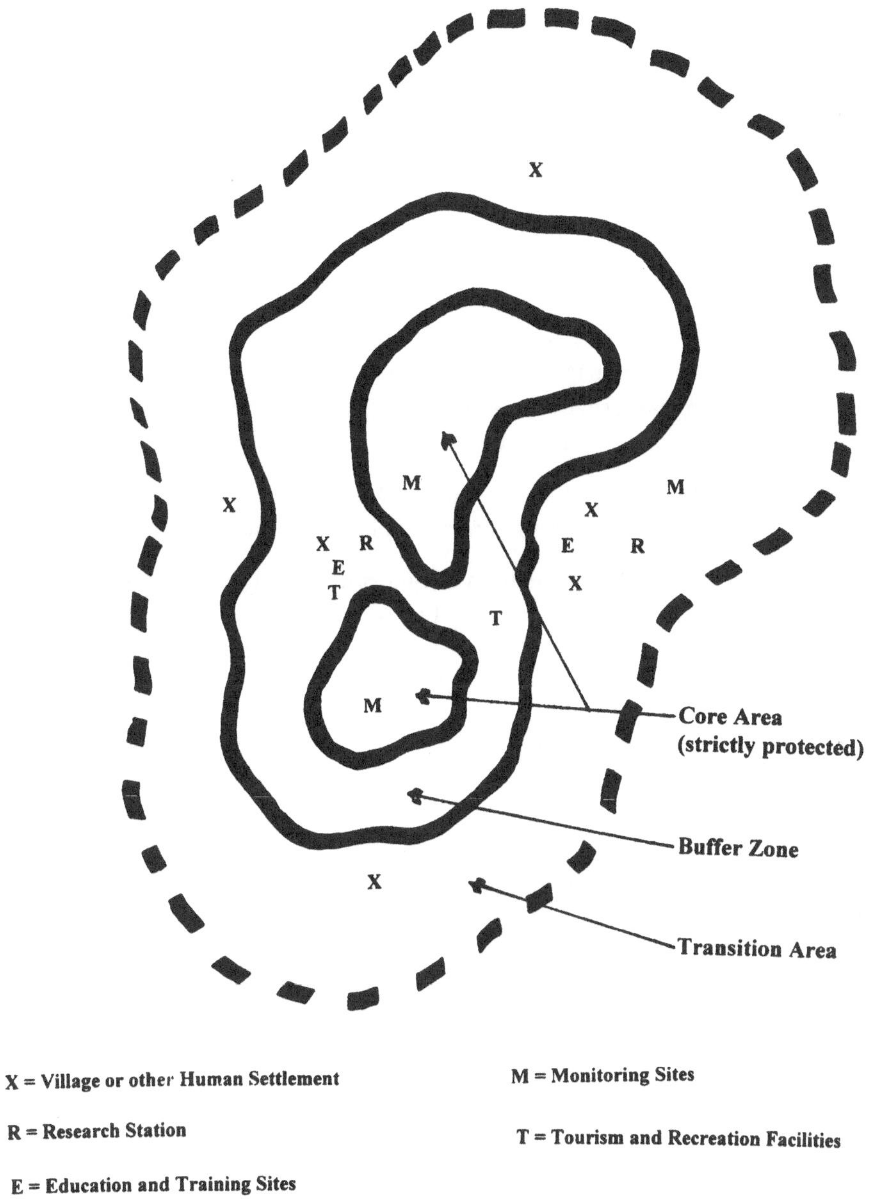

Figure 2.2 Generalized biosphere reserve design, showing various land use classifications. E, education and training sites; M, monitoring sites; R research station; T, tourism and recreation facilities; X, village or other human settlement. (Adapted from Batisse, 1982.)

vanish. Involving local communities in all stages of protected area planning and management is increasingly recognized by conservationists as an essential foundation for successful biosphere reserve management.

Yet in urging the adoption of biosphere reserve principles, we must also acknowledge that they have proven difficult to implement in practice. According to Michael Wells and Katrina Brandon [240]

> There are now more than 300 biosphere reserves worldwide, but little attention has been paid to promoting development in the buffer zones or transition areas. One reason is that most biosphere reserves have been superimposed on existing parks and reserves. The agencies responsible for managing these areas have usually lacked the resources, inclination, or ability to modify their management approach....[T]he change of status to a biosphere reserve has usually been in name only, with little change in emphasis or management philosophy.

Wells and Brandon found that, among a wide range of integrated conservation and development projects attempting to implement buffer zones

> Current usage of the buffer zone term is inconsistent and overlooks practical problems to such a serious extent that the utility of the concept has been jeopardized...two recent reviews have concluded that there are in fact very few – if any – convincing working [buffer zone] models.

In the light of these problems, resource managers will have to develop new management arrangements with communities in and around protected areas, instead of merely implementing conventional approaches. How might they approach this task? In a critical review of buffer zone management, Jeffrey Sayer [205] found that the most successful buffer zone projects

> ...have not been short-term aid projects but initiatives taken by local community groups or resource managers who have made creative attempts to solve the day-to-day problems which they faced (cited in 52).

In an effort to spark such creativity amongst land managers and conservation professionals, this manual will introduce themes around which managers and researchers can focus their efforts to better understand local priorities and facilitate collaborative management of useful plants. Many situations exist where local residents, guided by traditional management, have been wise custodians of plant resources. These experiences reflect the potential of local communities to organize and regulate their own land and resource use patterns to maintain valuable species and habitats.

However, it is important to remember that not all local communities are inherently interested in plant conservation or share all the priorities of resource managers. Traditional management practices are no guarantee that any particular group or community will be effective stewards of all resources in all situations. This is particularly true when cultural lifestyles, belief systems, and subsistence patterns are undergoing rapid flux [236]. Anthropologist John Cordell [54] has played devil's advocate with many truisms in reference to conservation and traditional tenure systems among indigenous peoples.

> Indigenous societies probably were and are neither significantly better nor worse than European societies at preserving their environments. Many species of plants and animals suffered extinction at the hands of Pacific island cultures, for example, before colonists ever sailed over the horizon. So the traditional tenure systems at issue here, which have come down through the ages, are not panaceas for environmental degradation; they are not formulas for maintaining communities in some ideal state of isolation and equilibrium with their lands. They are not the ancient customs that once existed. Indigenous tenure systems operate today in vastly different economic and political settings than in the past.

Resolution of this issue, Cordell suggests, will be a crucial point of debate between indigenous peoples and conservationists, whom he holds

> are on the threshold of a new era as far as achieving better communication and coordinating their respective aims. The challenge for all involved in the various ends of this intriguing business is to blend a sensitivity to social justice concerns with a critical perspective on the environmental management potentials of indigenous tenure and other living customs today.

Cordell is correct to focus upon social justice as a key issue in natural resource management; at some level, all conservation and resource management activities are political acts. While protected area managers may not always find it easy to promote an image of neutrality, research projects and management initiatives often do carry the potential to foster equity and strengthen community stewardship of local resources. Anthropologist Nancy Peluso [164], in reference to forest management in Indonesia, suggests that forestry officials and protected area managers should see themselves in the catalytic role of redistributing wealth from wild plant resources out of the hands of a select few, instead of merely protecting the resources.

In making such redistribution, priority must be given to cultures whose traditions and livelihood are rooted in a particular place, and who in many cases have created and maintained the landscapes that managers presently seek to protect. Plant conservation efforts that recognize and include the cultural contexts in which plants are valued, utilized, and sustained, will be more collaborative in spirit and ultimately more successful. The methods and concepts presented in the following chapters will be most useful if managers approach plant conservation with the following questions in mind [231].

- How do local residents use plant resources, and why?
- What effects do these uses have on natural habitats, ecological communities, and the health of plant resources in nearby wildlands?
- What changes in the conditions of local residents promote patterns of use that deplete or conserve plant resources and local wildlands?
- What incentive might residents have to conserve local plant resources?

3

Working with local communities

Figure 3.1 Seri Indian community members attending a workshop in Bahia Kino, Mexico, to discuss management options for ironwood (*Olneya tesota*). The Seri fashion and sell ironwood carvings, one of their primary sources of monetary income. The workshop was prompted by Seri concerns about ironwood overharvesting due to competition from Mexican carvers and land clearance to promote exotic pasture grass. (Photograph: J. Tuxill.)

After stating the case for professional–local collaboration in managing plants and other natural resources, the question becomes: how can managers and conservationists promote collaborative management on the ground? In this chapter, we examine a range of approaches – some drawn from the social sciences, others from community development and natural resource case studies – that can make management of plant resources more collaborative. Much of this material comes from the work of organizations and institutions working to increase the participation of rural residents in conservation and development. In practice, there is a range of resource management approaches that commonly receive the catch-all label **participatory** (Table 3.1). Yet only a subset of these strategies truly engage local residents in ways that are not only participatory but likely to produce long-term improvement in peoples' management of plants and other natural resources [170].

Table 3.1 A gradient of participation in conservation and development projects (reprinted by kind permission from Pimbert and Pretty, 1995)

Typology	Components of each type
Passive participation	People participate by being told what is going to happen or what has already happened. It is unilateral announcement by an administration or by project management; people's responses are not taken into account. The information being shared belongs only to external professionals
Participation in information-giving	People participate by answering questions posed by extractive researchers and project managers using questionnaire surveys or similar approaches. People do not have the opportunity to influence proceedings, as the findings of the research are neither shared nor checked for accuracy
Participation by consultation	People participate by being consulted, and external agents listen to views. These external agents define both problems and solutions, and may modify these in light of people's responses. Such a consultative process does not concede any share in decision-making and professionals are under no obligation to take on board people's views
Participation by material incentives	People participate by providing resources, for example labor, in return for food, cash, or other material incentives. Much *in situ* research and bioprospecting falls into this category, as rural people provide the resources but are not involved in the experimentation or the learning process. This is commonly called participation, yet people have no stake in prolonging activities when the incentives end
Functional participation	People participate by forming groups to meet predetermined objectives related to the project, which can involve the development or promotion of externally initiated social organization. Such involvement tends to occur not at the early stages of project cycles or planning, but rather after major decisions have been made. These institutions tend to depend on external initiators and facilitators, but may become self-dependent

Table 3.1 *(continued)*

Typology	Components of each type
Interactive participation	People participate in joint analysis, which leads to action plans and the formation of new local groups or the strengthening of existing ones. It tends to involve interdisciplinary methodologies that seek multiple perspectives and make use of systematic and structured learning processes. These groups take control over local decisions, such that people have a stake in maintaining structures or practices
Self-mobilization	People participate by taking initiatives independent of external institutions to change systems. Such self-initiated mobilization and collective action may or may not challenge existing distributions of wealth and power

While moving land management practices towards interactive engagement with local residents is an important goal, it is not always an easy step for land managers. In part, this is because true collaboration transfers a certain degree of decision-making authority to local communities. For some professional land managers, particularly those from countries where community land rights are broadly respected, sharing authority may not pose a problem. In some cases, these are the unsung individuals doing the most to advance collaborative approaches. Other managers may hold sole authority for managing protected areas, but recognize that past confrontational and coercive interactions with local residents have not worked. Collaborative approaches will open up lines of communication, and give managers insight into the perspectives and realities of their neighbors. Better communication and understanding may not automatically solve conflicts over management goals for wildlands, plants and other natural resources, but will enable points of conflict to be identified, mutually acknowledged, and potentially negotiated for resolution among all participants.

3.1 Documenting and understanding local perspectives

Land managers and conservationists can learn much about how local communities manage natural resources from the work of anthropologists, rural development experts and grassroots activists from universities, community development organizations and similar institutions. Yet they should also be prepared to work and learn directly with rural residents. This section presents a set of interdisciplinary information-gathering methods that can illuminate residents' priorities, constraints and perspectives on managing plants and other natural resources.

These tools have a number of names, but are most commonly referred to as **participatory rural appraisal (PRA)** or **rapid rural assessment (RRA)** methods. Many were designed to be implemented over a relatively short time period by small, interdisciplinary teams of investigators, extension workers and/or land managers working alongside local residents [118]. While we often discuss these techniques as if they would be implemented by an appraisal team, they can also be used effectively by one or two individuals meeting with local residents over an extended period.

In general, rural appraisal tools are designed to collect information – both broad and specific – on socioeconomic relationships, land management patterns and traditional ecological knowledge of residents in a given community and its environment (Table 3.2). PRA and RRA are popular approaches that are currently much in vogue among conservation and grassroots development organizations. Although the rural appraisal tool-kit offers a constructive approach to beginning work with local residents, it is also important to note its limitations [162].

Most rural appraisal tools are short-cut methods of more elaborate and often statistically valid techniques developed by researchers in a variety of different fields, such as anthropology (interviewing), geography (map-making) and ecology (transects). The initial outline which rural appraisal tools can give of a site or issue will invariably need to be colored in, verified and modified over time. This can be done through more intensive and quantitative research techniques, and by simply maintaining extended communication and information exchange with local residents. While rural appraisal can speed up the initial learning process, it will probably not reveal all local subtleties that are important for the conservation and management of plant resources over the long term. In addition, rural

Table 3.2 Types of rural appraisals (reproduced by kind permission from Freudenberger and Gueye, 1990)

Approach	Objectives	Results	Examples
Exploratory	Assemble general information on a broadly defined issue or problem	Preliminary findings or hypotheses; indicate focal points for further appraisal work	Document the agroecosystems of a region. Identify plant resource management issues of concern to local residents
Topical	Deepen understanding of a particular issue or problem	Specific hypotheses; recommendations for collaborative management	Survey resident's use of wild plants as food. Document why farmers in a village maintain landraces
Planning	Involve local residents in defining and implementing collaborative management activities	Program or plan of activities researched and designed by local residents	Participatory exercise to outline a management plan for local plant resources
Evaluative	Evaluate the impacts or results of a project, program or management effort	Revision of project goals in light of experiences. Readjustment of management approaches	Evaluate impacts from changes in resource use regulations. Review efforts to promote indigenous tree species in agroforestry systems

appraisal tools do not automatically generate an effective working arrangement between resource managers and local communities. Like any tool, their ultimate value for collaborative management of plant resources depends heavily on the underlying skills, outlook and priorities of the individuals doing the fieldwork.

3.1.1 Documenting local perspectives: before you begin

Preparation for any collaborative appraisal starts well before the first interview or initial community map-making session. There are several steps that managers and researchers should take in advance to ensure they make the most of their time in the field with local residents:

- Undertake background research.
- Make appropriate community introductions.
- Take note of potential biases.

(a) Undertaking background research

In many cases, secondary data sources on the field site or region already exist and can provide valuable background information. Previous academic research may have documented regional environmental, social and historical conditions. Government reports and maps (often called **gray** or unpublished literature) can also be extremely useful. By allocating time for background research, conservationists can avoid duplication and refine the focus of their field investigations.

(b) Making appropriate community introductions

Any collaborative effort must establish good rapport with community members prior to beginning local fieldwork. The manner in which conservationists are first introduced to local residents can be particularly influential, and it is always helpful to reflect beforehand how you might be perceived or what you might represent in local eyes. For instance, if you are introduced to a community through state officials, local people may regard you as a law enforcer, thus inhibiting discussion of current land use patterns and other important information [164]. This situation occurred in Tamaulipas, Mexico, when establishment of the Rancho el Cielo Biosphere Reserve terminated extractive logging in the region by large commercial forestry industries. To provide economic alternatives, managers wanted to encourage small-scale cottage industries based on propagating ornamental plants. However, local residents initially confused researchers and managers with law enforcement officials. Until they were assured otherwise, they feared being arrested for any subsistence hunting or plant gathering [218]. When beginning work in a community, it is wise first to hold formal and informal meetings with community members to discuss the goals and methods of the appraisal. Such meetings should stress that the appraisal participants are there to learn from community members, and that the information gained during the fieldwork

will lay a foundation for collaborative work on resource management issues. Experienced rural appraisal practitioners also suggest being careful not to raise community expectations that investigations will lead directly to development projects or economic benefits in the community.

Finally, whenever conservationists visit new households or villages during the course of gathering information, it is important that they re-introduce themselves and review the purpose of the survey with local assistants. Such actions will go a long way towards maintaining an open, productive atmosphere among all collaborators.

(c) Taking note of potential biases

In any sort of participatory documentation, it is important to recognize, and where possible minimize, any biases in the information that is collected. The most important way to minimize bias is by obtaining multiple perspectives on a single event or piece of information. This process of seeking different viewpoints is also called **triangulating** the information you receive, because often at least three viewpoints are used to home in on an issue or point of information.

Multiple perspectives can be gained in at least three ways. First, wherever possible, assemble an appraisal team with interdisciplinary backgrounds and complementary strengths. For instance, the team should not be composed entirely of biological scientists (better to include social scientists and community development experts), nor should it consist only of academic researchers (better to include extension workers and local residents who have served as resource managers). In addition, the team should include individuals fluent in the local language, and both men and women should be involved to minimize gender biases. By having diverse and complementary perspectives, you will be more likely to gain multiple views of a single event.

Second, bias is minimized by talking and interacting with local residents from a range of different backgrounds and therefore from different viewpoints: men, women, young, old, wealthy, poor, various ethnicities, different professions and so on. Participants should be a cross-section of the community and include men and women from various age and socioeconomic groups. Particularly in communities where men's and women's roles are sharply delineated, it is essential that the perspectives of both genders be solicited for a balanced understanding of natural resource issues (Box 3.1). Additionally, it is important to identify and

Box 3.1 Hearing both men's and women's perspectives is important

Anthropologist Margaret Jacobsohn [114] has worked extensively with rural communities in Namibia on conservation issues. In the village of Purros, resident pastoralists had begun selling baskets woven from palm leaves to tourists at a newly developed safari lodge nearby – such baskets

are also used to hold milk when villagers milk their cattle. Concerned about the potential impacts this new trade could have on nearby populations of the favored basketry palm, *Hyphaene ventricosa*, Jacobsohn and a fellow conservationist sat down with Purros villagers to discuss the danger of overharvesting the palms.

At the meeting, one of the oldest men in the village reminded the assembly that under traditional harvesting practices, only one or two fronds were removed from a palm during each growing season. All individuals present agreed this traditional approach would be an appropriate way to manage the palm harvest. The village elder promised that he would personally monitor use of the palms.

Yet several months later, more than 20 denuded palms near the village were clearly suffering from overharvest. The elder who monitored the palms blamed the village women: 'They are too stupid or too lazy to do what we had agreed. They are killing all the nearby trees.'

Subsequent conversations with the women of Purros revealed that the palm resource had long been under their control, since women gathered the fronds and wove them into baskets. In addition, when milk from the cattle (men's property) was poured into the palm baskets for serving, the milk became the property of women, who then controlled its distribution. From the women's perspective, placing a male elder in charge of monitoring use of the palms represented an attempt to transfer the palm resource – and by extension, the milk supply – to the men of the village. Women were protesting this potential loss of power by deliberately overharvesting the palms.

Balance was restored in Purros when the male elder agreed to give up his monitoring role, and the women were consulted on what they felt was the most appropriate way to manage the palms. The women's ideas proved workable, for as Jacobsohn recounts, 'hundreds of baskets later, the palms still thrive in the Purros area' [114].

include particular individuals who have greater knowledge of the appraisal subject than most community members. For plant resources, such locally recognized specialists may include herbalists, healers, expert farmers, midwives, sacred grove custodians and others.

A third way to control bias is to employ a variety of information-gathering tools to get at any one particular issue. All of the tools outlined in the pages that follow can yield different perspectives on a single issue or group of related issues of concern to the community. Information that is not revealed by an interview (perhaps something embarrassing to the participant) may become evident in a timeline or seasonal calendar.

Finally, the importance of regularly reviewing information from fieldwork and cross-checking for potential biases cannot be overestimated. As rural appraisal practitioners Karen S. Freudenberger and Bara Gueye [82] suggest:

> Teams should schedule regular pauses in their field activities and stop to ask themselves: 'Have we been cross checking all the information we're getting using different techniques and various sources?'

3.1.2 Participatory information-gathering tools: interviews

Perhaps the most straightforward way to increase collaboration between local residents and conservationists is simply to get both groups to sit down together frequently, talk openly and listen carefully to what each other has to say. **Interviews** or **oral histories** are tools to start such communication moving forward. Interviews can be conducted with groups or with individuals. **Group interviews** and collective oral histories of places have the advantage of presenting multiple perspectives from a single community all at once. The resulting interplay between residents during discussions often reveals valuable information on local social arrangements.

However, group meetings, particularly when held in a formal setting, can be easily dominated by a few prominent or assertive individuals, such as local political leaders or village elders. **Individual** or **focus-group interviews** allow more reticent community members, particularly women and the poor, to express their views more openly [62]. Focus-group interviews target particular **subgroups** of people within a community who have specialized knowledge, such as village herbalists or women farmers. Members of a subgroup may be linked by gender, social status, occupation, ethnicity, age, or similar variables.

The question of how you decide whom to interview is an important one. If you are seeking a cross-section of the community, or if you plan to do any statistical analysis of the information you obtain from interviews, then it is important that you interview a **random** selection of residents. In discussing this point, ethnobotanist Gary Martin [133] writes:

> Choosing your informants randomly implies that all members of the population have an equal statistical chance of being included, ensuring that your sample will not be biased in favor of any particular social group ... This is particularly important in situations where the dominant social group – men, elders, wealthy, or highly educated people – tends to be the first to come forward to be interviewed.

One way to obtain a random sample is to assign a number to each house in the community where you are working, write each number down on a slip of paper and then pull the numbers out of a hat until you have your desired sample size. Or you can pick houses and flip a coin at each one – heads you interview a family member, tails you pass the household over – until your desired sample size is achieved [133].

Of course in other situations you may choose people to interview through entirely non-random ways, particularly when you are interested in talking with people about specialized local knowledge. Knowledge of useful plants often falls into this category, as does community history (talk to the elders) and local folklore (visit the story-tellers). Selecting interview subjects in a non-random way does not automatically mean you are collecting bad or inaccurate data, but you should have your eyes open for possible biases and for opportunities to confirm information independently with other individuals. In addition, it is always useful to record a set of general data about an interviewee that allows you to place them in a social context within the community [133]. Such information often includes a person's age, gender, ethnicity, education level (years of schooling, literacy), family status (married or not, number of children) and family history (recent migrant or long-term resident).

Interviews can have a range of different formats. One possibility is **open-ended interviews**, which generally are relatively free-ranging discussions covering many different aspects of local life. You can think of open-ended interviews as a relatively informal way to learn from local residents when you are new to an area, or have not previously held discussions with community members.

Once you become more familiar with local issues and begin to interact with residents, **semi-structured interviews** can help you to gain detailed information on particular topics. Semi-structured interviews are less rigid than surveys – rather than having a formal list of pre-written questions, you work from a checklist of topics that serves as a reminder of what you are seeking to learn. The checklist will usually evolve during the course of fieldwork, as initial questions become resolved and unanticipated but important issues come to light. The step-by-step protocol presented in Box 3.2 for interviews follows a semi-structured approach – though most of the steps are relevant for other interview formats as well.

Box 3.2 A protocol for semi-structured interviews

Beforehand

1. Materials needed:
   – pen and small notepad for taking down notes;
   – (optional) portable recorder with blank cassette tapes and batteries.
2. Prepare a table stating your desired interview objectives, the topics you would like to discuss, and which people seem the most appropriate individuals to interview (Table 3.3). For instance, do you want to hear from a cross-section of the community, or will it be more efficient to interview only those subgroups of people who have specialized knowledge about plants? You can also write down a few sample questions for each topic as a guide for how to phrase questions (see Box 3.3).

3. Decide on a target sample size of interviews you would like to conduct, as well as a method for selecting participants. Also think about how you might use other tools described in this section – such as maps or ranking exercises – in interviews, and prepare for these accordingly.

Table 3.3 A sample outline guide for planning semi-structured interviews. In this case, the interviews are designed to provide socio-economic information on the artesanal use of *cocobolo* (*Dalbergia retusa*), a legume tree native to Panama and other Central American countries. Note that the interviews are expected to cover more ground than just the questions listed, which serve primarily to guide the interviews as they unfold

Objective	Whom to interview	Sample questions
To learn about the supply of *cocobolo*	artisans cocobolo suppliers	– In what kind of forests do you find *cocobolo*? – How much time do you spend collecting? How far do you have to travel? – When do you collect? How many trips per year? How much do you collect at one time? – How much *cocobolo* do you collect yourself and how much do you buy from others? –From what parts of the tree do you take wood? Why?
To learn how *cocobolo* carvings are produced	artisans	– How did you learn to carve? Where did you learn? – How do you decide what to carve? – How much time do you invest in each carving? – What tools do you use to carve? – How often do carvings break? What do you do with the leftover wood? –How much do you sell your carvings for? To whom do you sell? How often do you sell? What amount of your income comes from carving? – What would you like to change about your carvings? – What problems do you encounter?
To learn how *cocobolo* is marketed	craft vendors artisans	– From whom do you buy carvings? – What is the range of prices you pay? What is the average price? – What designs sell best? What market trends have you observed? – Who are your typical clientele? What percentages are repeat customers? How much do they buy? – What recommendations do you give to artesans?

During

4. Pay attention to small details that may help participants to feel relaxed and at ease. Choose a familiar, comfortable spot for the interview, such as under a shady tree at the edge of a person's yard. Sit at the same level and position as the participants. Explain your purpose or goals for the interview, and assure participants their responses will remain confidential.

5. The most reliable way to record information is with pencil and notebook. Tape recorders and cameras may make some participants uncomfortable – ask carefully before using them, particularly if you are a newcomer to the community.

6. It can be challenging to write notes, listen intently to participants, and simultaneously judge the subtle body language or expressions people give with responses. Be prepared to practice this skill beforehand, perhaps in practice interviews with friends or during initial open-ended interviews. If you are part of an appraisal team, you and the other team members may want to interview in pairs – at least for group interviews with the community – so that one person can take notes while the other concentrates on participants' responses and body language.

7. Work from your topic list, but don't be in a hurry to complete it. Probe any responses from participants that are not clear or that raise interesting points not anticipated in your checklist. Also be aware when participants are reluctant to answer a question directly – a sign that you are touching on a sensitive issue – or are becoming restless, an indication that the interview has probably gone on long enough and the participant wants to move on to other tasks.

8. At the interview's end, thank participants for their time and contributions. Leave a copy or summary of your notes with participants, if they so request. This is also an opportunity to ask whether they have any questions for you.

Follow-up

9. Review your interview notes later in the day, both to fill in details not written down and to identify potential biases. Did all participants appear to express their views freely, or was the conversation dominated by particular individuals? Also, how might the information you gained influence your interviewing sampling approach? Did you obtain unexpected information that suggests the presence of issues or subgroups of individuals in the communities whom you had not recognized before, but should now investigate further?

Sources: [62, 82, 133].

Interviews can also be structured by joining residents in their daily activities (referred to by anthropologists as **participant observation**) or visiting specific sites that provide a backdrop for the issues the interviewer is seeking to learn about. One way to interview local residents about wild plant use is to take different individuals or small groups of the same age to a location on community lands or within a protected area (for instance, a standardized 1 ha plot) and ask them the names, uses and rates of regeneration of any plants they know. This technique depends upon residents feeling comfortable enough to volunteer information to you, rather than you directing them to particular plants for which you presume uses. Involving people from different age groups or user groups can offer insight into whether different subsets of community members value the same plants in different ways. Such techniques can also be employed to investigate the intergenerational retention or loss of traditional ecological knowledge

The types of questions that you ask can heavily influence the effectiveness of the interviewing process. Box 3.3 offers a number of suggestions for formulating effective interview questions.

Box 3.3 How to ask the right questions

There are several pitfalls which frequently trap interviewers when asking questions.

Closed-ended questions

Questions answerable with a simple 'yes' or 'no' can make it difficult to get a discussion underway. It is better to phrase such questions in an open-ended way that requires participants to expand upon a subject. For example:

> When you find wild plants growing in your garden, do you always weed them out?
> could be better stated as:
> *What do you do with wild plants when you find them in your garden?*
>
> Are you going to prepare beans for dinner tonight?
> *What will you prepare for dinner tonight?'*

One way to avoid closed-ended questions and other problematic questions is to adopt an interviewing strategy which uses questions that probe and explore to illuminate local complexities. These questions begin with:

- Who?
- What?
- Why?
- Where?
- How?

Leading questions

These questions imply a desired response, and are guaranteed to bias the information gained during an interview.

Neem seed make good pesticide, don't they?

What local products can be used as pesticides?'

Implicit assumptions

When a question contains an implicit assumption, it can be confusing for the respondent.

Will you collect chicle tree latex or palm fronds in the forest today?

If the respondents plan to collect allspice berries, how are they supposed to respond? They might correct the interviewer but, out of politeness, they might not. A better question is:

What plants do you plan to collect in the forest today?

Vague questions

When a question has more than one interpretation, the respondent may have a very different perception of what is being asked, leading to confusion.

Is it hard to pound millet?

The interviewer could have dehulling in mind, while the respondent thinks she means grinding to flour. Or the interviewer may be referring to the physical exertion of pounding, while the respondent is thinking about the time pounding requires. A better question might be:

What steps are there in pounding millet?'

Source: adapted with permission from Karen S. Freudenberger and Bara Gueye [82].

3.1.3 Making maps

One of the inherent problems in establishing and managing protected areas is that prior land uses and current traditional harvests are not necessarily apparent to professional land managers. Individuals, families or communities who have utilized the land in the past may be reluctant to admit directly their ongoing uses, for fear of arrest by rangers or wardens. Historic uses may be even more difficult to discern, perhaps because woodcutters or farmers no longer live in the area, or because their influence on the vegetation is so subtle that land managers do not identify the area as one experiencing human use.

One potential remedy for this problem is to engage long-term residents in drawing **vernacular maps** which identify former trails, harvesting sites, hunting blinds, fields, fallows and homesteads that have been abandoned. 'Vernacular' means 'in the common way of talking' and such maps are useful tools for both local residents and professional managers. In biosphere reserves where traditional uses of the biota are not prohibited, vernacular maps may help community members formally establish usufruct rights or land tenure to particular parcels. They may also allow land managers to recognize patches of hitherto unnoticed secondary vegetation amidst primary forest, or to identify buildings, agricultural terraces or oases worthy of preservation as traditional cultural properties. Finally, local residents may be enlisted to help map routes and entries used by visiting poachers, wildcat loggers, clandestine miners and others whose activities are inconsistent with the desires of the community and protected area managers [214].

Vernacular maps vary from ephemeral sketches in the sand or on a blackboard, to individuals' hand-drawn paper maps of their plant gathering grounds, or a community's carefully drafted map of their homeland, derived from consultation with numerous elders (Figure 3.2). Vernacular maps may emphasize culturally salient features that fail to appear on government-generated topographic maps, such as a row of transplanted edible prickly pear cacti (*Opuntia*) that marks the territorial boundaries of two tribes [211]. Some vernacular maps are derived entirely from individuals' perception of their homeland, in which the imagined representation of spatial relations and selection of salient features is put at the discretion of the individual. Anthropologists refer to these as **cognitive maps**, although a drawn map is actually an imperfect by-product of such a mental image. Through open-ended questions such as 'Can you make us a map of where you live and gather resources?' or 'Can you draw a map of where you trek?', participants are encouraged to hand-draw maps or sketches of what they feel is relevant to their own lives. Projects in Central America, Canada, India and other regions have encouraged local residents to draw such maps, outlining areas where hunting, fishing, gardening, gathering and ceremonies have taken place – in relation to well known topographic features [174, 176].

Cognitive maps are often combinations of sketches of horizons, coastlines and prominent features – half map, half drawing – with perspective shifting from vertical to horizontal. They may not necessarily be centered on the spot where the mapper is working or living at the time. For example, the Seri Indians along the Sea of Cortez in Mexico often center their maps on the place where their parents buried their placentae. While Seri maps are sparse in human-constructed features, they are accurate in details of coastline topography and mountain horizons [103]. Most importantly, composites of these maps identify the 20th-century domain of Seri subsistence activities that affect the distribution and abundance of native organisms. Although there are plant and animal populations which the

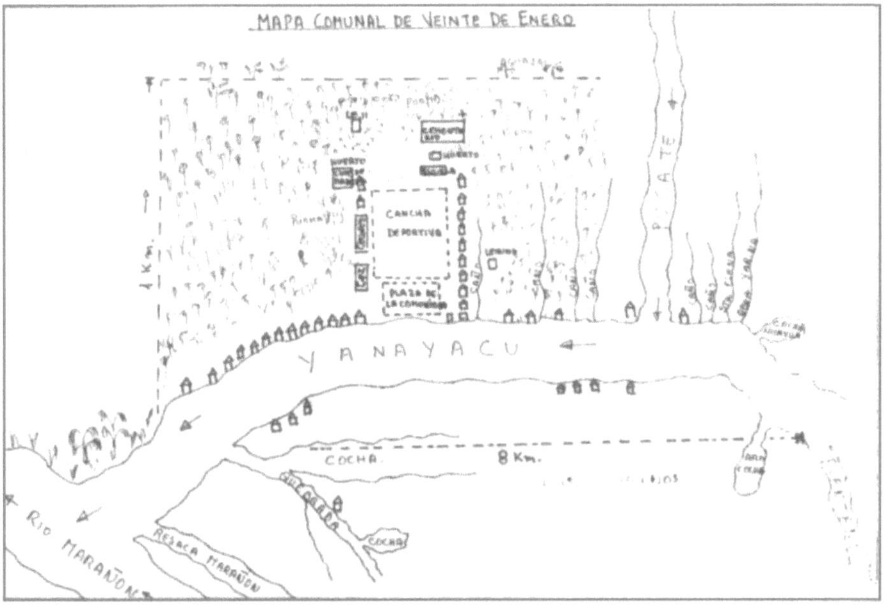

Figure 3.2 A sketch map of the Peruvian village of Veinte de Enero, prepared by community members during a participatory rural appraisal sponsored by Fundación para la Conservación de la Naturaleza, a Peruvian non-governmental organization. (Reprinted by kind permission of The Nature Conservancy.)

Seri may have historically introduced to additional islands across a much wider area than their present range [254], contemporary Seri have no memory of these translocations.

Geographic information on traditional land and plant uses can also be documented through group meetings where community members jointly prepare maps of their surrounding lands. Figure 3.2 is this kind of map. Such maps can be sketched directly on the ground as well as on paper; the ideal location for a ground map is a large, flat area where villagers can spread out comfortably, perhaps a school courtyard or under a tree outside a communal building. Useful materials for map construction may include twigs, seeds, stones, leaves and soil of variable color, all of which can represent different features. Local schoolchildren can help gather these materials when you are preparing to construct a map.

In other cases, you can present community members with **base maps** of the region and ask participants to sketch in boundaries or color in circumscribed use areas with pencils or marking pens. Base maps are maps that have been formally prepared by professional cartographers, usually working in government-sponsored mapping agencies. You can use such maps with local residents by overlaying clear plastic overhead sheets atop base maps of topography, soils, political boundaries, or even atop satellite photos. Participants can then draw in locally recognized features on the overheads. These overlays may begin as provisional discussion pieces, without any strictures regarding precision or scale. However, be prepared to explain the base maps in detail, for they are often confusing to participants who have never worked with them before.

A general protocol for vernacular map-making is outlined in Box 3.4. In all instances where vernacular maps are prepared, outside facilitators should try to interfere as little as possible and allow locals to debate and decide what the map should include. Often, the very absence of locally utilized place names on a government-drafted topographic map stimulates discussion. In one such situation, a predominantly non-Indian community in northern Mexico publicly acknowledged that the formerly unnamed 'Canyon de los Papagos' was historically a floodwater farming site of displaced Native Americans (O'odham, or Papagos).

Box 3.4 A community map-making protocol

Beforehand

1. Materials needed:
   - for ground maps, use sticks, stones, seeds, leaves, soil of variable color, straw and other local items;
   - poster-sized sheets of blank newsprint or other clean, white paper;
   - marking pens or colored pencils;
   - tape for affixing the corners of paper sheets to a table, wall or other flat surface;
   - formal survey maps of the region, if available;
   - clear plastic sheets for overlays (make sure your pens will mark on the plastic);
   - camera and film for photographing the finished map;
   - waterproof map sealant (for humid climates).
2. Review your purpose for conducting a map exercise, the information you would like to gather, the mapping methods you will use, and the range of people you would like to participate in the exercise.
3. If you are making a ground map, gather the necessary materials before starting. If you are making the map as a group exercise, choose a central location where everyone will be comfortable.

During

4. Begin by explaining to participants why you are interested in having them make a map of the area. After this initial explanation, it is best to let participants proceed with minimal interference from yourself or other outsiders. Be sure to note the dynamics among participants as they discuss what to include or exclude from the map.

5. Formal base maps can be confusing for people who have not used them before. When using base maps, explain them thoroughly to participants before they begin working with overlays. Point out rivers, hills, towns and other local features that appear on the base map, so that participants can orient themselves. Then, to judge how they are reading the map, ask participants to point out or give their interpretation of other features.
6. If a large group is present, consider breaking people up into subgroups (such as men and women) to prepare maps independently. The subgroup results can then be discussed by the entire group and combined into a master map. While this process will likely take longer than mapping as one group, it will get more individuals involved and reduce the likelihood of one or two people dominating the exercise.

Follow-up

7. Discuss the map with participants. What local features did not appear on the map? What items are you surprised to see included? Does the map communicate the information you were originally seeking to gather?
8. Sketch the finished ground map on a piece of paper and photograph it for a permanent record. Make sure the community or group of participants receives a version of the map produced – we recommend leaving the original map with participants and taking a copy for your own records. If you are working in a humid climate, waterproof the map with sealant.
9. Paper or plastic overlay maps can be carried into the field for **ground-truthing**, or direct confirmation of map features. If a GPS unit is available (discussed in the following pages), you can use it in the field to establish formal coordinates for map features and also to quantify map distances.
10. Emphasize to local map caretakers the importance of storing maps in a cool, dry place out of direct sunlight, where the map will not fade or become tarnished.

Sources: [62, 82, 175].

It is common for a completed map to generate many subsequent discussions. Maps constructed on the ground can be re-sketched on a sheet of paper and taken to the field for verification. In subsequent discussions you may want to have residents draw more specialized maps detailing individual land holdings, locations of key plant resources and other subjects of interest. Such specialized maps are particularly useful when discussing proposed changes in management plans for an area or evaluating the impacts of management activities [175]. For instance, Dianne Rocheleau and colleagues [193] asked residents in the Zambrana-Chacuey region of the Dominican Republic to map the plant resources of their individual land holdings, as part of a project to evaluate the social and ecological changes resulting from an agroforestry initiative (Figure 3.3). This initiative, sponsored by both the national forestry agency and an

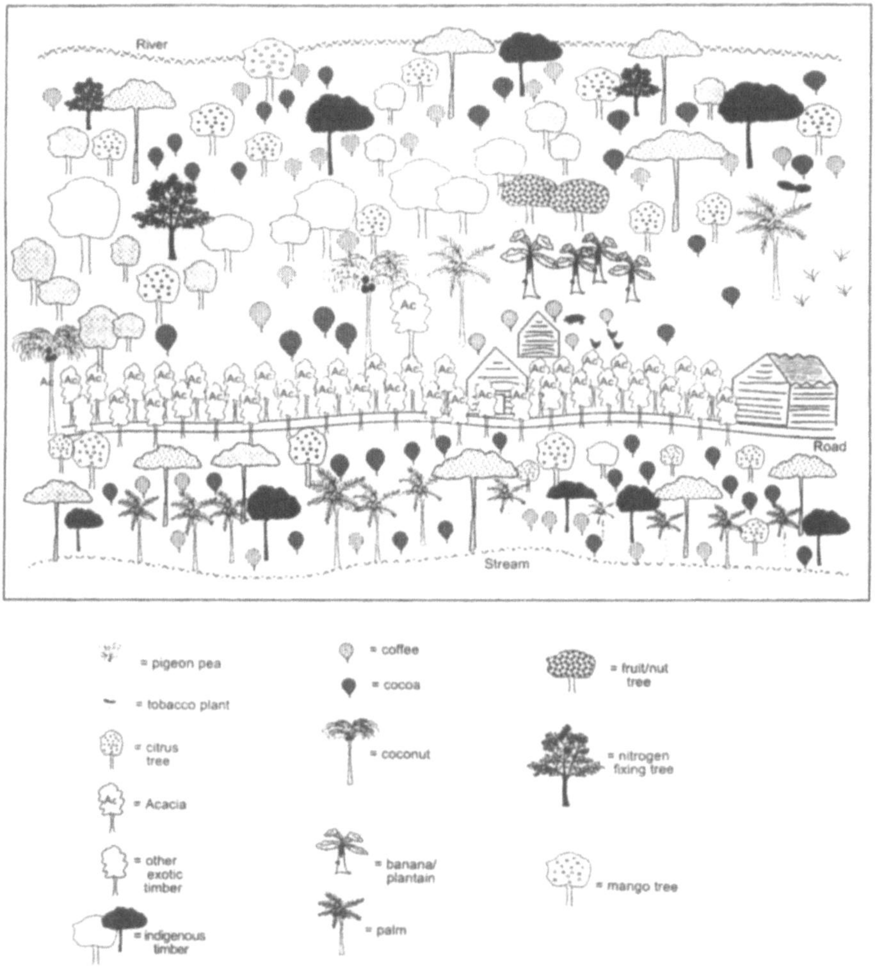

Figure 3.3 Recopied sketch map of the farm of Hernando and Marta Perez near the region of Zambrana-Chacuey, in the Dominican Republic. This map emphasizes the kinds of trees present on the Perez farm; it was prepared as part of a regional agroforestry research project that examined social and ecological patterns of land use changes related to a social forestry program promoting an exotic fast-growing timber tree, *Acacia mangium*. By focusing on a single farm, the map achieves a greater level of detail than the community map in Figure 3.2. In addition, it demonstrates how recopying and adding a key to a vernacular map can help to make it easier to interpret for people unfamiliar with the area. (Reproduce by kind permission from Rocheleau *et al.*, 1996.)

international NGO, promoted a fast-growing tree, *Acacia mangium*, for small-landowner timber production in the region.

It is also becoming increasingly common to use hand-drawn vernacular maps in the field to identify important locations such as plant-gathering sites, and then to reference these sites for later inclusion in a **geographic information system (GIS)** database [176]. Geographic information systems are computer programs that can store, analyze and display spatial information, especially in the form of computerized maps. GIS are particularly useful for combining different kinds of data – for instance, on soils, rainfall, or the location of culturally prominent sites – and then analyzing the spatial correlations that result [202]. Certain new GIS programs can be run on relatively small computers yet still provide high-quality maps and analyses (see Appendix A).

Global positioning system (GPS) devices enable individuals to map features in the field accurately, the results of which can then be transferred into a GIS for analysis. GPS units are portable hand-held devices that gather signals from a global satellite network to calculate accurately the latitude and longitude coordinates and altitude of the site where the device is placed. Many GPS units calculate positions that are accurate to within 100 m (though vertical altitude readings are usually less accurate than horizontal longitude and latitude readings). Some units also are capable of storing the coordinates of surveyed points and then transferring or **downloading** the data directly into a GIS program.

How can these technologies help the managers and stewards of rare plant resources? There are many dynamic applications of GIS map analyses which have advantages over investing more time in mapping resources on multiple paper maps [202]. For example, point data regarding the latitude, longitude, altitude, soils, slope and aspect from known locations of threatened plants can be entered into a GIS database and then used to generate **surfaces** or maps of potential plant population distributions within the larger area covered by a map. This potential area of distribution can then be searched or ground-truthed in the field, and areas where the plants are absent can be entered into the database to improve knowledge of the species' range. In this way, additional point data can be used to refine or test a distributional model, ultimately saving botanists many hours of search time in the field.

To help to monitor traditional plant harvesting, data from vernacular maps identifying harvesting areas can be overlaid on plant abundance data (section 5.4). In this way, you can discern whether the condition of plant populations in harvested areas differs from that in areas free of harvesting, or whether a combination of factors – say, harvesting plus frost susceptibility in cold air drainages – have dovetailed to trigger plant declines.

Another recent combination of vernacular mapping and GIS mapping has been described by Rhodora Gonzalez [93] from her work in Neguev, Costa Rica. Gonzalez first had local inhabitants describe their own folk classification of soils and soil suitability for different crops. She then correlated the folk classification with formal soil taxonomy. With this information, Gonzalez developed a GIS database of computer-generated maps of various scales delineating areas potentially suitable for particular crops based on local knowledge of plant/soil relationships. However, it is not always easy to establish one-to-one correlations of western soil or plant taxa, with indigenous soil or plant taxa [108]. In such cases, western-trained scientists must rethink their own assumptions about what technical data are useful to local inhabitants, relative to inhabitants' own rules of thumb [93].

3.1.4 *Transect diagrams*

Transect diagrams are similar to maps in that both portray landscape features, but because transects are drawn from a different perspective they can provide additional information. Transects typically indicate topography and major zones of land use, along with important resources or activities in each zone (Figure 3.4). A rural appraisal team working in India found that transect diagrams were particularly fruitful when drawn from an elevated spot, such as a rock outcrop or the rooftop of a house [175]. To ground-truth the transect, or if an elevated viewpoint is not available, villagers and team members follow a direct path across the landscape, noting, sketching and verifying features along the way (Box 3.5).

	River	Fields	Village, Bamboo and Fruit Trees	Degraded Sal Forests	Protected Sal Forests (7-8 Years)	Village
Management Problems	--River bank erosion. --Flooding.	--Rainfed, marginal farming, small holdings.	--Heavy dependence on fuelwood headloading for income. --Lack of unity on effective FPC functioning.	--Overcutting and overgrazing of sal coppice growth. --Soil erosion.	--Rapid sal regeneration under effective FPC protection, but continued threats from neighboring villages. --Inadequate intermediate product flows.	--Lack of access to low cost rice (*padi*). --High interest rates on loans.
Management Opportunities	--Bamboo planting on banks. --Bank stabilization with low cover shrub vegetations.	--Fodder and fast growing fuelwood species (trees and shrubs) on bunds.	--Develop small NTFP industries: leaf plates, mushroom cultivation, bamboo products.	--Form FPC and establish access controls. --NTFP species, enrichment planting, target headloading families specifically.	--Encourage FPC's of neighboring villages to act cooperatively. --Allow more rotational felling and thinning with NTFP enrichment planting.	--Storage facility for rice pounding. --Establishment of FPC bank account and credit system.

Figure 3.4 Example of a transect diagram. (From Poffenberger *et al.*, 1992, reproduced by kind permission of Society for Promotion of Wastelands Development.)

Box 3.5 *A transect diagram protocol*

Beforehand

1. Materials needed:
   - large sheets or pads of paper (but not so large as to be cumbersome when walking in the field);
   - colored pencils or waterproof marking pens;

– notebook and pencil for recording additional information;
– (optional) GPS unit for logging coordinates of transect.

2. Review your purpose for conducting a transect exercise, the specific information you seek to gather, and the people whom you would like to have participate.

During

3. Explain to participants that you would like to make a diagram showing the different habitats or kinds of land uses near their house or community. Ask them if they know a spot along a ridgetop, rooftop, rock outcrop, or other elevated location where a broad view might be obtained. If such a spot is not available, the next step may be skipped.
4. Working from the view in front of you, ask participants to sketch landscape patterns along a straight-line path towards a prominent feature in the distance. The sketch may include land-use zones, habitat types, topographic divisions, or other features deemed important by participants. Discuss each transect zone with participants, and have them fill in details on management problems, dominant vegetation, important plant resources and other relevant information.
5. With participants, walk as direct a line as possible across the area to be diagrammed. If the transect was already sketched from a vantage point, follow the line-of-sight chosen for the sketch and ground-truth information as you go. If step 4 was skipped, any direct course chosen by participants can be taken, and different zones or habitats sketched and discussed along the way.

Follow-up

6. Discuss the completed diagram with participants, particularly to elaborate on points for which you could only make quick, brief notes in the field.
7. You may want to re-draw the transect on another piece of paper and arrange the information about different zones below the diagram, as in Figure 3.4. Do this step with participants present, so that they can fill in any gaps that may remain in the information.
8. Leave the original transect diagrams with participants, making copies for yourself and other interested outsiders.

Sources: [62, 82, 175].

3.1.5 Marking time: timelines, seasonal calendars and activity schedules

There are a number of ways to quantify temporal and historical patterns at a survey site. Local residents meeting as a group can construct a timeline of the prominent events in a community's history (Figure 3.5). For this activity, the views and memories of village elders and other long-time residents can be particularly valuable. Land management patterns and changing environmental conditions at a site can be tracked over time using this same technique. Changing vegetation and habitat conditions can be qualified using a historical transect, representing the condition of a given site transect at various points in time (Figure 3.6). Box 3.6 provides a protocol for implementing timelines and historical transects.

c. 1700	Samkedji founded; first water point identified.
early 1800s	Mbardali's reign of terrror; Mbardali killed.
mid 1800s	Islam introduced.
c. 1905/6	Serer–Wolof battle at Diobas. French assistance Wolof → defeat of Serer. Beginning of Wolof domination.
c. 1933	First Fulani settlers arrive; whole village could fit in one bus.
1939/1945	Second World War; great hunger period.
c. 1955	Last year rice grown in Samkedji.
1956/7	Locust invasion, great hunger period.
1964	School constructed in village.
1965	Drought, hunger period.
1969	Last year Sanyo (late millet) grown.
1974	Arrival of grand marabout. Drought, hunger period, food aid.
1982	Last government provision of peanut seed. Drop in peanut cultivation.
1983	Drought; hunger period.
1985	Canadian gardening and water project; drought, hunger.
1986	Construction of seed-cereal bank; milling machine installed.
1987	Drought, hunger period.

Figure 3.5 Historical profile from the village of Samkedji, Senegal. (Reproduced by kind permission from Freudenberger and Gueye, 1990.)

Having local residents draw seasonal calendars is a powerful way to get at seasonal changes in local labor demands, crop plantings and harvests, availability of wild plant resources, local weather patterns, prices for locally harvested commodities and many other factors relevant for *in situ* plant conservation (Figure 3.7). Seasonal calendars specific to a single factor can be combined to produce a composite calendar (Figure 3.8), which can yield important insights into relationships among different factors and perhaps explain evident trends [82]. Activity schedules can express daily activity patterns and be useful in quantifying

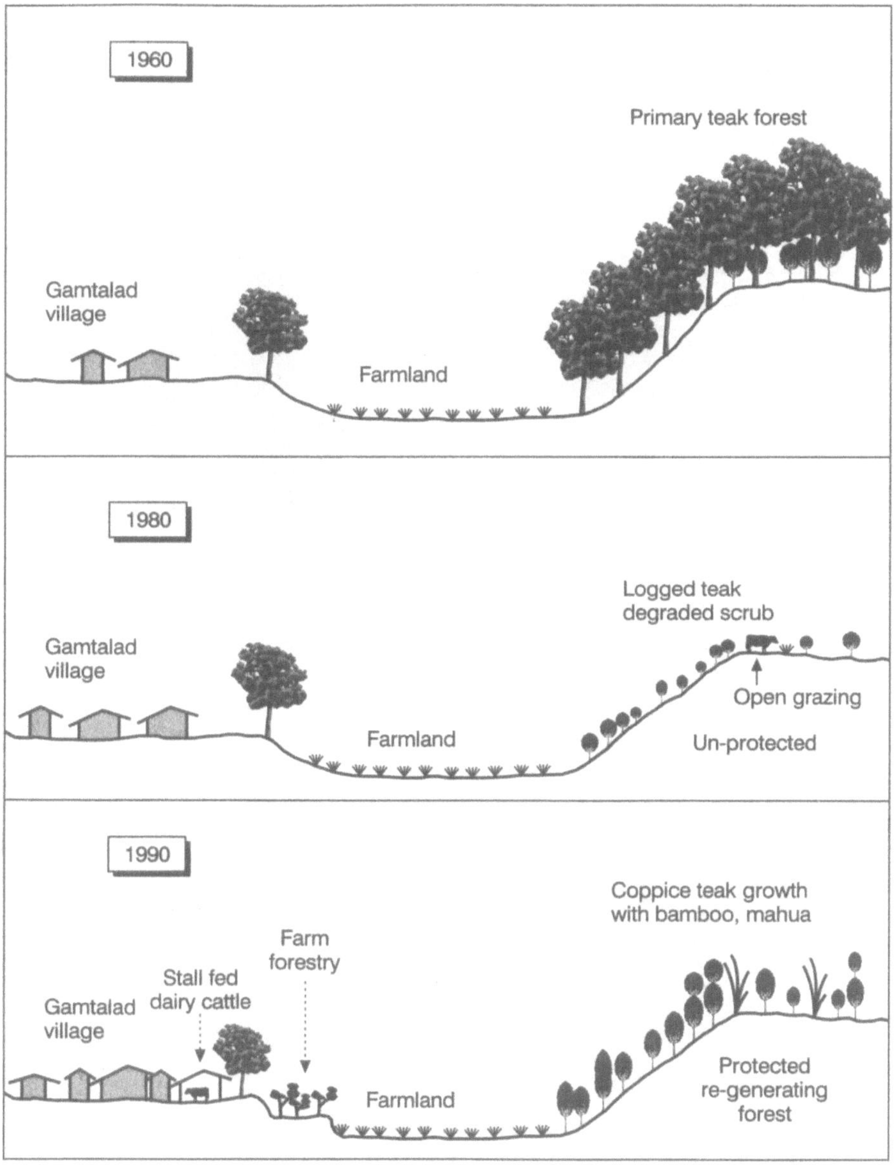

Figure 3.6 Historical transect from the village of Gamtalao, India. (From Poffenberger *et al.*, 1992, reproduced by kind permission of Society for Promotion of Wastelands Development.)

Box 3.6 Compiling timelines and historical transects

Beforehand

1. Materials needed:
   - large paper sheets;
   - colored pencils or marking pens;
   - (optional) blackboard and chalk may substitute for paper.
2. Review secondary literature or talk with individuals familiar with the region of study to learn about prominent past historical events, and to become familiar with local time units – not all societies use western weeks, months and years.
3. Community elders and other long-time residents are particularly important contributors to a timeline or historical transect. Contact such individuals in advance and make a special effort to ensure their participation.

During

4. Sit down in a comfortable spot with participants and explain that you would like to know about the history of their community or location. You may want to leave the topic broad at first, letting residents communicate what they see as the most important past events and changes. Subsequently, you can ask them about more specific trends, such as the local abundance of plant resources over time.
5. To develop a timeline, try working back from the present, asking questions like: 'What changes in your farm have occurred in the last 10 years? In the 50 years prior to that?' If years or other time units do not provide a good breakdown, you can also ask questions in terms of generations ('What changes in farming have you seen in your lifetime? What changes did your father talk about?') or in reference to major past events ('How did the big drought affect your farm?'). In this way you can help participants to assemble a timeline even without precise dates.
6. For a historical transect, ask participants first to sketch the current state of the area or habitat under discussion, and then work backwards from this baseline. Let participants decide on the past periods to include in the drawing – usually three to four will suffice. A few well-placed questions can help focus the diagram ('How many houses did the village contain 20 years ago? How much denser was the forest in your father's time?').

Follow-up

7. Review the completed diagram with participants to clarify any apparent contradictions or unclear spots. Also double-check timelines and transects with other knowledgeable elders who may not have participated in the original compilation of the diagrams.
8. Leave the original diagrams with participants as a community or family history, taking a copy for yourself and other interested outsiders.

Sources: [62, 82, 175].

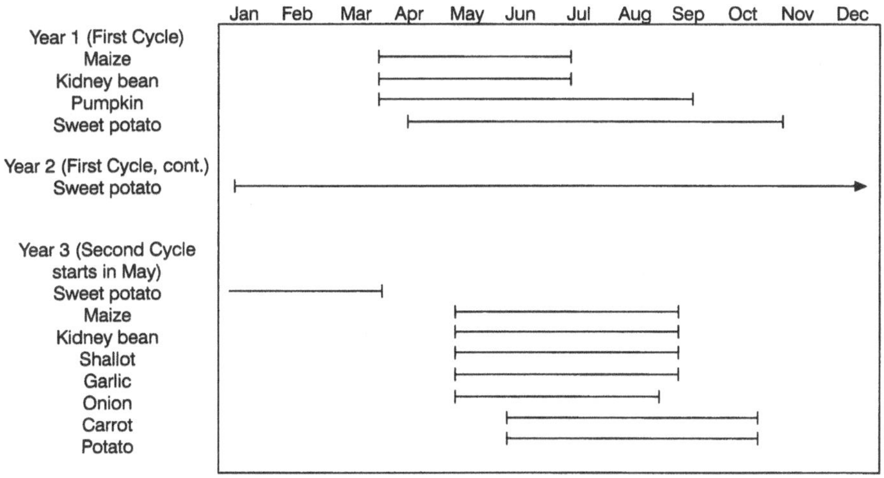

Figure 3.7 A seasonal cropping calendar from the villages of Iray and Sururey, Indonesia. (From Prain *et al.*, 1995, reproduced by kind permission of CAB International.)

labor outputs (Figure 3.9). Box 3.7 presents a protocol for seasonal calendars and activity schedules.

Freudenberger and Gueye [82] mention two additional guidelines to keep in mind regarding calendars. First, when drawing a linear calendar as in Figure 3.8, use an 18-month time scale to ensure that events or seasons spanning the start or end of a year do not get cut in half on the diagram (this is particularly useful for the composite calendar). Second, although all societies have ways of measuring time, it is not necessarily a western year divided up into months. As with timelines, be prepared to construct calendars and activity schedules using local units of time.

3.1.6 Making comparisons

One limitation of rural appraisal is the difficulty of quantifying the information

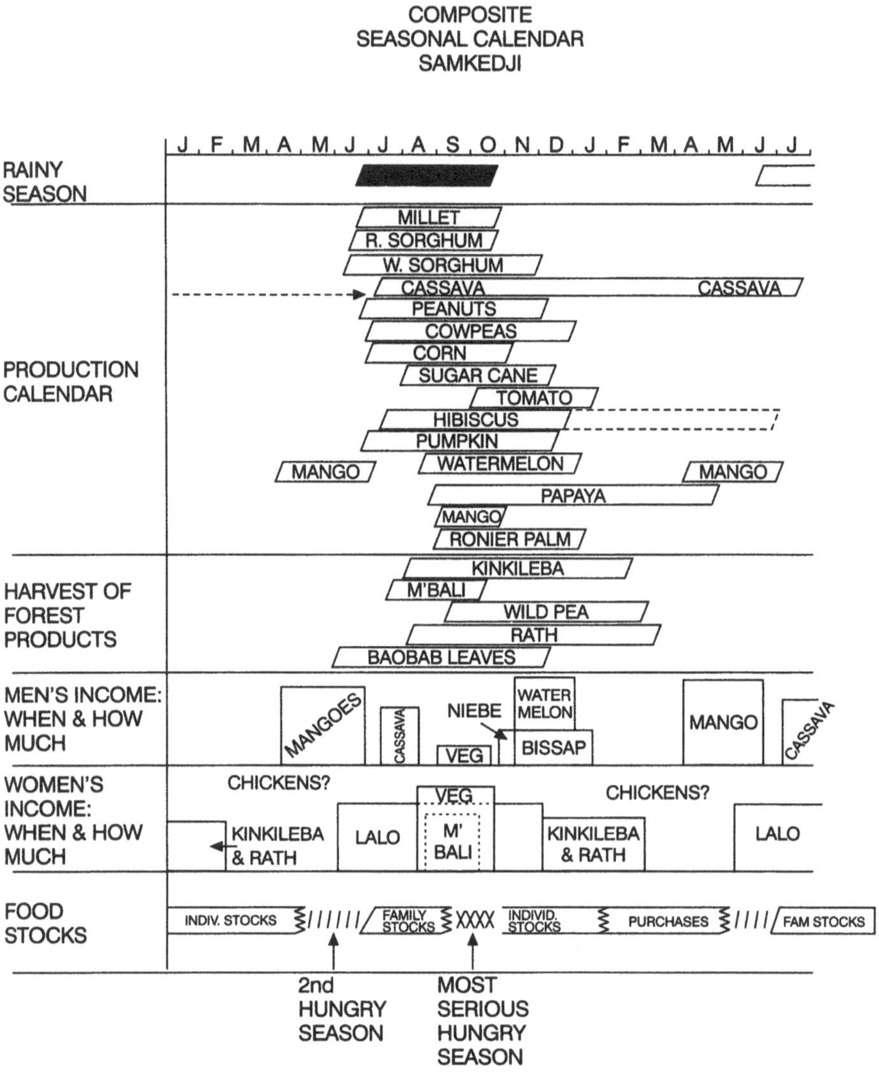

Figure 3.8 Composite calendar from the village of Samkedji, Senegal. (Reproduced by kind permission from Freudenberger and Gueye, 1990.)

supplied by local residents. Standardized survey questionnaires administered to a random sample of residents are one way to get around this problem and obtain information that is both quantitative and suitable for statistical analysis. However, for an initial picture of a site, precise numbers may be less important than relative comparisons between different items or factors, such as the relative contribution of different wild plants to meeting household needs for food, fuelwood, medicines and so on.

4 AM	
5 AM	Wake up: Sweep.
6 AM	Fetch water (Milking)
7 AM	Breakfast cooking. To forest for fuelwood.
8 AM	↓
9 AM	
10 AM	Walk roundtrip with headload to Ukai dam and sell
11 AM	↓
12 Noon	
1 PM	Return home: wash clothes.
2 PM	Grind grain.
3 PM	(Milking)
4 PM	Fetch drinking water.
5 PM	Free time with children.
6 PM	Begin cooking dinner.
7 PM	
8 PM	
9 PM	Sleep.

Figure 3.9 Daily activity schedule of women from Moti Pipal, Limbi Panchayat, India. (From Poffenberger *et al.*, 1992, reproduced by kind permission of Society for Promotion of Wastelands Development.)

Box 3.7 Constructing seasonal calendars and activity schedules

Beforehand

1. Materials needed:
   - large sheets of paper;
   - colored pencils or marking pens;
   - tape for affixing the corners of sheets to a table or wall;
   - beans, pebbles, or other counters;
   - small stones (for indicating months or other time units when calendars are sketched on the ground); camera and film.
2. Review your purpose for compiling an activity schedule or seasonal calendar, the specific information you seek to gather, and the people whom you would like to have participate.

3. Calendars and especially activity schedules tend to work best when done with community subgroups (women, men, poorer families, craft artisans, etc.) or individuals. Daily schedules and work calendars often vary significantly according to people's gender or occupation, and these tools can produce confusing information if subgroups are not separated out beforehand.
4. Learn the local time units by reviewing secondary sources or talking with local residents before beginning construction of schedules or calendars.

During

5. Find a comfortable spot and explain to participants that you are interested in people's daily schedules, how their work or use of plants varies season by season, or other time-based trends.
6. Seasonal calendars can be drawn at least two ways – as a circular or as a linear calendar. Participants can draw them on paper sheets or sketch them directly on the ground, using stones as time unit markers and beans or seeds as counters to indicate intensity of activity at different times.
7. You can initiate calendars by writing down the months of the year (or other local time units) as headers on a sheet, and then ask participants to indicate how crops, wild plant harvests, labor demands and other variables vary seasonally. Building a calendar one crop, plant product, or labor demand at a time will reduce potential confusion at this stage.
8. Do activity schedules by asking participants to write down how they spend a typical day and night. You can indicate hourly divisions in a header and have participants fill in activities alongside (Figure 3.9), or simply let participants prepare the entire schedule.

Follow-up

9. Calendars prepared for different variables can be combined and redrawn as a single composite calendar (Figure 3.8). Composite calendars often clarify temporal connections between different variables or activities, and are an excellent tool for follow-up discussions.
10. When calendars are sketched on the ground, be sure to copy them on paper and photograph the completed calendar for a permanent record.
11. Leave the original calendar or schedule with the community, making copies for yourself and other interested outsiders.

Sources: [62, 82, 175].

Anthropologists and other social scientists have developed a host of participatory field methods for having local assistants compare, rank and sort different variables. These methods include preference rankings, card sortings and triad comparisons, all of which are described in detail in the People and Plants manual *Ethnobotany* [133] by Gary J. Martin. Here we will focus on one adaptable ranking method that utilizes a matrix table (Figure 3.10).

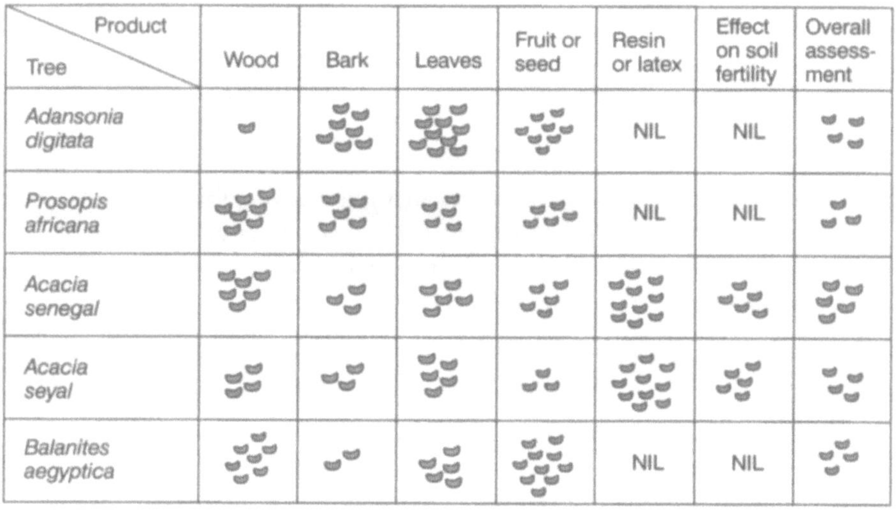

Figure 3.10 Example of a matrix comparing the products obtained from several tree species common in semi-arid regions of east Africa. Each cell is ranked on a scale from 0 to 10, using beans as counters, to indicate the relative value of each product produced by each tree species. (Adapted from Rocheleau *et al.*, 1989, and Freudenberger and Gueye, 1990.)

Matrix ranking involves setting up a cross-referenced chart and having community members record the relative importance of each item on the chart. In the above example, various plant products or benefits are placed along one side or **axis** of the matrix, and prominent local tree species (already identified through interviews or other methods) are arranged along the other axis. It is important to involve community participants in the identification and selection of axis categories as well as in the actual ranking of items. Once selected, participants place a desired number of beans, stones, or other counters – out of a given total number of counters available – on each matrix square to represent the relative importance of each product for each corresponding plant. A protocol for preparing a matrix with community members is outlined in Box 3.8.

This protocol covers both standard and **pairwise ranking** matrices. A pairwise or **preference** matrix (Figure 3.11) requires participants to compare several different categories or items directly and choose one over the other according to

Box 3.8 Preparing and using matrices

Beforehand

1. Materials needed:
   - large sheets of paper;
   - colored pencils or marking pens;
   - dry beans or uniformly-sized pebbles for counters.
2. Review your purpose for compiling a matrix, the specific information you would like to gather, and the people whom you want to participate.

During

3. Sit down with participants and explain that you would like them to compare the importance of different trees, crops, time demands, or a similar topic. Often, you can use matrix comparisons to follow up on initial discussions or interviews.
4. On a sheet or clean, smooth patch of ground, sketch a large empty rectangle with two blank margins or **axes** on the top and left sides.
5. Discuss with participants the items or **categories** to be compared, and the **characteristics** that will be used to compare them. Write the categories down the side axis, and the characteristics along the top axis. For example, in Figure 3.10 the tree species are the categories, while their products or benefits are the characteristics. Once you have the axes filled out, draw in the internal walls of the matrix to create a lattice of cells for comparisons. The exact number of cells will of course vary with the number of categories and characteristics you are comparing.
6. Think about how you can represent categories and characteristics by symbols and pictures as well as word labels – this will make the matrix more accessible to non-literate participants. For instance, when different plants are being compared, samples of each one can be placed along the axis.
7. Once the axes are complete, think about how you will move through the matrix to fill in the cells – this can affect how participants make comparisons. For instance, in Figure 3.10, moving **across** one entire horizontal row at a time emphasizes comparing the different uses of a single tree species. Working **down** a vertical column emphasizes the comparative value of each tree species for one particular use. Try to fill out the matrix in a pattern that makes the most sense for the particular information you are trying to gather.
8. To begin comparisons, hand out beans or other counters to participants

to place in each matrix cell. There are several ways to help to standardize participants' comparisons. One is to distribute a pre-counted number of beans per matrix, such as 100 beans for a 20-cell matrix. Or try setting a range for responses, such as between 0 and 5 beans per cell. However, in some cases participants may ask to add more counters, the better to represent a comparison, and they should not be dissuaded from doing so.

9. Participants can work together and come to agreement on each score to award to a cell, or each participant can offer his or her own score one by one and the results are then averaged at the end across all participants. The first approach may be faster in some cases, but the latter has the advantage of involving all participants.

10. Be aware that many community subgroups (particularly men and women) may score items differently when comparing them. If you are working with a mixed group, try dividing each category into two rows, one for each gender.

11. To create a **pairwise matrix**, list the categories to be compared along **both** axes, and black out half the cells as shown in Figure 3.11. Each of the remaining cells represents a comparison between the two categories that correspond to that cell. Have the participants work along the matrix, deciding which of the two categories they rank higher according to the characteristic being used. Fill in each cell with the preferred item.

12. Once the pairwise matrix has been completed, tally up the number of times each category appears in the cells. Rank the categories in descending order based on these tallies.

Follow-up

13. Discuss the completed matrix with participants to learn why they scored the categories as they did. Insight into the reasons why participants make particular decisions is some of the most valuable information that matrices can provide.

14. If the matrix is done on the ground, photograph it and make a paper copy. Leave the original matrix with participants, but don't forget to take a copy for yourself.

Sources: [82, 158, 195].

a standard criterion, such as overall usefulness, amount of labor required to grow or harvest, or similar characteristic. Pairwise matrices are useful in situations where you are trying to rank several items directly, for instance when establishing priority species for *in situ* management. As Figure 3.11 indicates, a pairwise matrix only has half the squares of a standard matrix. Each square represents a comparison between two categories, and participants must fill in the square with the category that is more useful, labor-demanding, etc. When the matrix is completed, each category will have been compared with every other category, and the total number of times a category appears indicates its overall score and rank.

Native agroforestry species	*Adansonia digitata*	*Prosopis africana*	*Acacia senegal*	*Acacia seyal*	*Balanites aegyptica*
Adansonia digitata		*Adansonia*	*Acacia senegal*	*Adansonia*	*Balanites*
Prosopis africana			*Acacia senegal*	*Acacia seyal*	*Balanites*
Acacia senegal				*Acacia senegal*	*Acacia senegal*
Acacia seyal					*Balanites*
Balanites aegyptica					

Tree species	Number of times preferred	Rank
Adansonia digitata	2	3
Prosopis africana	0	5
Acacia senegal	4	1
Acacia seyal	1	4
Balanites aegyptica	3	2

Figure 3.11 Example of a pairwise ranking matrix. In this case, five tree species common in semi-arid regions of east Africa are ranked to assess their value as potential species for agroforestry. (Adapted from Odour Noah *et al.*, 1991; information on trees from Rocheleau *et al.*, 1989.)

One advantage of using matrices for ranking is that they can store a large amount of information. You can bring a completed matrix to subsequent discussions and ask why participants gave categories the rankings that they did. As a result, matrices can generate copious discussion and be very useful for analysis.

3.1.7 Compiling the results of fieldwork

Analyzing information and reporting results is how you document what a team or group has accomplished, the knowledge they have gained and the importance of their findings for a particular situation – say, for the management of plant resources in the buffer zone of a biosphere reserve. To make sure that results get compiled, analyzed and communicated, it is best to do these steps directly after the field appraisals. Much of the interpretation of findings ideally takes place as the fieldwork unfolds, but inevitably there will be summary analyses to compile. Make sure these are done as a group by all members of the appraisal team, while still in the field. One option for initial analysis is to prepare preliminary analysis reports for different aspects of the fieldwork, which list objectives, main findings (emphasizing visual displays of results) and suggested follow-up activities [175].

Doing initial analyses in the field ensures you can present results and analyses to the community and other local participants in a summary meeting, to give them a chance to provide feedback on what the appraisal team has learned. This kind of summary event is very important, for it returns information to the community and allows the appraisal team to cross-check findings before preparing the final, formal documentation.

Final documentation of rural appraisal can take a variety of forms, but usually involves, at a minimum, a formal written report. Writing up results can take place as a team, with members concentrating on their areas of expertise, or writing can be delegated to specific team members. If the latter takes place, all team members should have an opportunity to review the report before it is completed and distributed. As you sit down to prepare any kind of formal report, there are several important questions that you should ask at the start.

(a) What should the report include?

Try to make your report as thorough as possible, given the time constraints or deadlines you may be under. A complete written report will usually include:

- an executive summary (one to two pages) detailing the survey purpose and key findings;
- survey objectives;
- methodologies and tools employed;
- general information about the site;
- specific findings;
- analysis of the situation;
- conclusions;
- a plan for follow-up work.

Visual materials are particularly important and compelling, so include copies of maps, transects, matrices and other visuals prepared during the fieldwork. Tables and graphs are efficient ways to summarize information – use them freely, but also remember they will just be a distraction if they are not clear and easily interpreted. As you write up your findings, analysis and recommendations, take the time to ask: Am I presenting this clearly? If I was reading this without any prior knowledge of the appraisal, would I understand it? How might I say it better?

(b) Who is your audience?

Think about who should know about your findings and how you can best present results to inform them. This question should be discussed in the field, before doing final analysis, so that local participants have the chance to indicate who **they** think should receive a report of the fieldwork. Often, the potential audience is quite diverse:

- community members who work directly with plant resources;
- other local residents, such as political leaders, who may be interested in your findings;
- extension workers and field representatives of government agencies and NGOs;
- bureaucrats and other desk managers within government institutions;
- research institutions;
- donor agencies;
- the general public.

Formal written reports are the expected form of communicating findings within an agency or institution, but will usually not reach people in many of the other categories. The important thing is to decide how you can best communicate information to your chosen audience and be prepared to use a variety of approaches to get your findings across. For instance, to disseminate findings on optimal management methods for certain plants used in artisanal products, the key recommendations in a formal report could be translated into a picture-oriented brochure for local residents. We talk more about how to communicate your work to a diverse audience in section 3.2.7.

(c) What do you want the report to accomplish?

Reports have the potential to catalyze action to resolve a problem and to provide guidance for follow-up activities, but all too often they are never used. Thus it is important to think not only about who your audience is, but how you would like them to respond to your findings. How well a report fulfils this role depends in part on how timely, persuasive and accessible it is for readers and will determine its ultimate usefulness.

3.2 Beyond appraisal: addressing local concerns in plant resource management

Rural appraisal methods can facilitate *in situ* plant conservation in many different situations. Protected area managers can work with local residents on mapping and transect exercises to clarify and monitor patterns of wild plant use in communities surrounding a nature reserve. Conservationists interested in learning about traditional gathering or agricultural practices in a region might use historical timelines, activity schedules and semi-structured interviews to gain an understanding of the constraints – and potential opportunities – facing communities who seek to maintain or recover subsistence traditions.

Documenting and acknowledging local residents' perspectives in this fashion is the first step toward collaborative management of plant resources. Collaborative management, when implemented fully, however, is more dynamic and comprehensive than just occasional field appraisals or exchange of ideas at regular intervals. This section explores approaches that managers and local residents can follow to build and strengthen collaborative management over time. Many of them apply the rural appraisal field methods just discussed.

3.2.1 Letting communities lead: participatory rural appraisal

When natural resource management is truly participatory, local residents have a strong voice in shaping the direction and emphasis of professional research activities, conservation planning and land management [170, 175]. Rural appraisal methods can help to give local residents this voice through a community development process known as **participatory rural appraisal (PRA)**.

PRA makes use of rural appraisal tools, but differs in the way appraisal results are applied. Following initial analysis and summary of fieldwork with the community, the PRA team and community leaders assemble a roster of problems and opportunities for action facing the community. In discussing and acting upon this roster, PRA offers local communities an opportunity to express, debate and pursue their own goals and objectives regarding land and natural resource management. Equally important, it can open up internal community dialogue beyond what previously existed [175]. PRA can also be applied to identify key local plant resources [58], which can subsequently be studied by professional scientists, resource managers and local residents, working in collaboration. A protocol for implementing PRA is presented in Box 3.9.

Box 3.9 Undertaking a participatory rural appraisal

As with other kinds of rural appraisal, PRA involves natural resource professionals – ideally an interdisciplinary team – collaborating with local residents to collect community-level information within a limited time frame. For PRA, however, professionals should strive to be facilitators rather than to direct discussion or just gather information. The general process goes as follows [158].

1. PRA begins, either through a direct request for problem-solving assistance from a community, or via the recommendation of extension workers, land managers, or other field staff familiar with the local situation.

2. Once initial interest in a PRA has been expressed, outside professionals and community leaders meet and clarify the issues to be investigated. PRA can be either topical or exploratory (see Table 3.2), depending on the nature of the problem identified and the goals that local leaders have for an appraisal. A date should be set for the field appraisal. Local counterparts for the appraisal work should be identified and selected.
3. Local counterparts and outside facilitators implement an appraisal session to collect information on the local community. At this stage, the tools discussed in section 3.1 are applied. Emphasis should be on having local participants collect information themselves as much as possible. Local individuals with some formal training, such as schoolteachers or extension agents, may be able to take the lead at this stage. While outsiders can still expect to demonstrate tools the first time they are used, the goal is to have residents take an active role in using them as the information-gathering proceeds.
4. Analyze and summarize the information obtained from the appraisal. Summarize information with composite maps, calendars and other diagrams that will help you to present results effectively. Even if you have to do much of the technical analysis and summary yourself, make sure you do it with local counterparts present and involved so that they see how you do the work. Delegate responsibility whenever possible.
5. Working from the appraisal findings, assemble a summary on paper of **problems** facing the community and **opportunities** for the community to take action and resolve the problems. Try to summarize problems and opportunities as clearly as possible.
6. Present the summary of problems and opportunities at an assembly of community members. If possible, have a local counterpart who has helped to gather and analyze the findings make the presentation.
7. Next it is the assembled community's turn to **rank** the identified problems and opportunities in order of priority, a step that is best done at the same meeting as the previous step.

If other problems or opportunities are raised besides those identified by the appraisal team, add these to the roster and rank them as well. Criteria useful for prioritizing opportunities include:

- **feasibility** (Can we do it? What steps will be required?)
- **productivity** (Will the results be worth the effort we put into it?)
- **sustainability** (Can we maintain it over time?)
- **equity** (Will it be fair to all community members?)
- **stability** (Will it make our lives more secure?)

8. If the assembled community members have trouble coming to agreement on how to rank problems and opportunities, try facilitating this step by using a pairwise ranking exercise (described in section 3.1.7).
9. Working with local counterparts and other community representatives, use the roster of ranked problems and opportunities to prepare a community resource management plan (CRMP). The CRMP should contain:
   - a **summary** of the **appraisal findings**;
   - a **copy** of the **problems and opportunities** facing the community, as ranked by meeting participants;
   - a **list** of the community's **strengths** and **weaknesses** that influence how the community may respond to problems and opportunities;
   - a **list** of steps that will be taken to **address problems** and **act upon opportunities**;
   - an **outline** of how progress on these steps will be **monitored** and **evaluated**;
   - a **list** of **who will be responsible** for implementing each step, including the monitoring and evaluation.

In style and practice, preparing a CRMP is similar to the planning process outlined in section 4.3 for managing plant resources. That discussion can guide you further in accomplishing this step.

10. Meet with community leaders and identify options for how to put the CRMP into action. In some cases, CRMPs have been used to design contracts formalizing the mutual responsibilities of community groups, government resource managers, NGOs and other institutions represented in a particular region. In several PRA experiences in Kenya, villages have used CRMPs to attract external support from government agencies or NGOs to help to accomplish desired community development goals [158]. Because they are generated and shaped directly by community members, CRMPs tend to have a high degree of on-site follow-up and to be appropriately attuned to local realities.

3.2.2 Participatory action research

It is not uncommon for local expert farmers, herbalists, gardeners, or wild plant harvesters to serve as assistants on formal research projects that are investigating the plants, farming practices, or ecological features of a region. This kind of research can be a useful adjunct to rural appraisal methods and offers many insights for managing plants and other natural resources. But it can produce even more benefits when local residents and conservation professionals work together to design, administer and interpret research on problems specifically identified by a local community, rather than exclusively by outside professionals.

This process, referred to by some community development practitioners as **participatory action research**, has several conservation applications [62]. It can be used to monitor and evaluate existing land use and agricultural practices, impacts on plants and other natural resources, and other environmental trends that local residents identify as important in their lives. In addition, it is a way to test out new technologies or changes in land use practices proposed by extension workers and other outside professionals. Such research can also serve as a training exercise to build the formal analytical skills of local residents.

Overall, participatory action research has three main characteristics that distinguish it from other research approaches [62]:

- **Local residents take a lead role in determining which problem, technology, or management method will be studied, and how, when and where it will be studied.** A group meeting between residents and conservation professionals is one potential forum for deciding the goals, formats and hypotheses to test through participatory action research. Such research can also be a follow-up action to address issues raised by a participatory rural appraisal.
- **The methods and tools used in the research must be readily understandable and able to be implemented by all individuals involved, local residents and scientists alike.** For instance, where local residents do not have high levels of literacy or mathematics training, participatory research methods initially may emphasize documentation and comparison of qualitative rather than quantitative trends. Working around these constraints can demand creativity and innovation on the part of professionals involved in participatory research. For example, professionals may devise a replicated sampling regime that can be statistically analyzed, but residents should take the lead on selecting sample sites and monitoring each sample. While this may limit to some degree the explanatory power of the investigations, the benefit is that the research process will be demystified for local residents and be more readily seen as a learning approach that everyone can apply on their own.
- **The research findings are analyzed, reported on and interpreted by local residents and outside professionals together.** The results and their implications should be discussed and shared among all parties. The goal of the research is to improve management of the plants or other natural resources of concern. Local adoption of plant management innovations identified through research is more likely to occur when residents are part of all stages of the learning process that generated the results. Even if professionals do the statistical analysis of locally collected data, the results should be interpreted in terms of the question, need, or hypothesis that the community originally posed. A protocol for PAR is presented in Box 3.10 and a field example in Box 3.11.

Box 3.10 Implementing participatory action research

1. Meet with community members to identify problems, new technologies, or new management practices that might be addressed through research or tested with the participation of local residents.
2. Once a problem or new practice to test has been identified, meet again with interested participants to design the research. Discussion should cover:
   - **what** variables need to be measured and monitored in the study;
   - **how** such information will be gathered;
   - **who** will do the measuring;
   - **what** sort of tools will be required.

Take plenty of time on this step (it is likely to involve more than a single meeting) and make sure local participants have a good grasp of the study methods and how to implement them. It may be necessary to hold training seminars or workshops to familiarize participants with systematic documentation methods before proceeding.

3. Implement the research plan with participants. Working together, select **control** sites, which are left undisturbed or maintained under typical management, and **experimental** sites where different interventions or techniques are tried. You may want to hold seminars where participants practice standardizing their selection of control and experimental plots or measuring sample variables.
4. Begin to collect data. Visit local data collectors throughout the study period to verify that they are following procedures correctly. Work with them further on methodology if necessary.
5. Once information has been collected for the desired length of time, sit down with participants to analyze and summarize the data. You can do this as a workshop, where participants review and practice the basic mathematical operations needed to analyze the information collected. Even if you have to take the lead on data analysis, do the work with the data collector present, so that they see the entire process.
6. Meet with participants to discuss:
   - what the results indicate;
   - which factors best explain the findings;
   - whether the results suggest making changes, continuing past practices, or doing further research.

It is good to discuss individual experiment results with each participant and also to gather all participants together to discuss their cumulative results. Group field visits to research sites that best demonstrate the results obtained can be a productive way to discuss and interpret the research.

7. Discuss with participants how best to share the research results with the greater community. Perhaps participants can present their results at a community meeting? Another possibility is to summarize findings as a pamphlet or management manual that can be distributed to people.

Sources: [62, 198].

Box 3.11 Participatory action research with Bolivian farmers

World Neighbors, a grassroots development organization, has employed participatory action research to increase food security among subsistence farmers in the Bolivian highland region of Potosí [198]. Organization field staff realized early on that the diverse microclimates evident among individual farms posed a problem for extending experiment station research. Agricultural research to optimize crop productivity would have to take place on-farm, and be directed by the farmers themselves. This approach quickly gained the enthusiasm of farmers, who met at monthly seminars to share their experiences with different varieties of potatoes and other staple crops, fertilizer applications, and legume plantings to increase soil fertility.

It was difficult to compare or apply individual farmers' findings, however, because few farmers recorded specific field data such as planting and harvesting dates, rainfall levels, yields and elevation. In part this was due to the oral tradition of local culture and low written literacy levels among farmers. Project staff realized that if they were to increase the productivity of local agricultural practices, farmers had to perform and document their experiments more systematically.

To this end, they began a pilot program involving 71 Potosí farmers. The farmers attended a series of seminars where they discussed the value of systematic documentation methods, including making replicate plots, assigning treatments randomly, establishing control plots, and other agronomic concepts. Farmers also got a chance to practice these techniques by planning and planting a small field trial on their own land. In addition, project staff organized group tours where farmers visited and compared each other's experiments.

The importance of documentation was reinforced when harvest time arrived. Seminars were arranged where farmers practiced harvesting the experimental plots, and observed directly how information recorded on rainfall, planting dates and other variables could explain differences in the results obtained. At these harvesting seminars, farmers were introduced to basic statistical concepts and practiced doing mathematical calculations on calculators and number tables. On their own initiative, a number of participants organized field days at which they shared their results with fellow farmers.

Over three years this farmer-directed research has yielded many agronomic insights, including identifying which varieties of eight staple crops work best at different elevations, and demonstrating how frost and drought damage to potatoes can be reduced by starting with a certain seed size. The experiments have also stimulated the desire of participants to improve their own technical skills. Many began requesting extra mathematical instruction at seminars, while others took up secondary education courses offered via radio. Farmers are clearly interested in the methodology behind the experiments. In the words of farmer Alejandro Mamani [198]:

> The experimental field trials have also influenced my wife. She always wants me to test new varieties and fertilizer practices prior to planting in larger demonstration trials. This has increased our family income.

This work has also produced *in situ* conservation benefits. Farmers' systematic documentation of cropping practices has improved communications with researchers from institutions such the International Potato Center (CIP). Researchers at the Center realized that the results farmers were getting with their diverse, traditional potato varieties were much better than results obtainable with a new potato variety CIP planned to promote among high-elevation farmers. Based on the findings of farmer experimenters, CIP cancelled the release of the new, more genetically uniform potato variety in the Potosi area.

3.2.3 Paraprofessional training

A related way to build local capacity for managing plant resources is to offer local residents the chance to gain formal conservation skills directly, through workshops or more extended training exercises. This training is most commonly referred to as **paraprofessional training**, since participants often learn many of the same skills employed by extension workers, resource managers and other professionals. Paraprofessional training may not necessarily identify or solve any specific problems, but instead may give local residents greater exposure to formal resource management approaches and improve their own ability to monitor and evaluate conservation activities.

Paraprofessional training can take a range of formats. Many conservation organizations sponsor workshops, seminars and short courses that introduce participants to a set of skills and procedures over a short – but intensive – time period. For instance, the Arizona-Sonora Desert Museum recently arranged a training workshop on water quality monitoring, in which one grassroots watershed protection group – Los Amigos del Río Santa Cruz – explained and demonstrated methods to a farmers' group – El Patronato Pro-Limpieza del Río Magdalena – in an adjacent watershed.

In other cases, paraprofessional training can be implemented as part of ongoing resource management activities. In the east Caribbean nation of Dominica, the Caribbean Natural Resources Institute has worked with a cooperative of small-scale lumber producers, Cottage Forest Industries (CFI), to make their timber harvesting more environmentally sound. Over a three-month period, three CFI sawyers were employed alongside one of us (JT) to inventory timber volumes on a government-owned forest stand the cooperative was interested in managing for lumber production. During the course of completing the inventory, the sawyers were trained in reading topographic maps, navigating with compasses, laying out transects and plots, measuring tree diameters, assessing regeneration and other forest inventory skills (Figure 3.12). While the inventory might have been completed more rapidly with a team of professional foresters, the ultimate goal of local self-reliance was better accomplished with an extended training program [228].

Figure 3.12 Reginald Williams and Handel Prosper, two members of a local timber harvesters' cooperative, training as paraforesters while conducting a forest inventory on the Caribbean island of Dominica. The training involved learning how to navigate with map and compass, lay out transect lines and establish inventory plots. (Photograph: J. Tuxill.)

3.2.4 Exchanges between communities facing similar challenges

Sometimes the best way for conservationists to assist local residents in managing plant resources is to put a community in contact with others who practice alternative management approaches, or who have maintained traditions or knowledge related to plants. By arranging and sponsoring exchanges between communities or local institutions, such as the aforementioned one between grassroots watershed protection groups, conservationists can foster improved resource management at the local level. This strategy is sometimes termed a **lateral technology transfer** to distinguish it from top-down approaches.

In 1994, conservation official Arturo Argueta began arranging exchanges between indigenous residents of biosphere reserves and other indigenous groups considering participation in proposed reserves in Mexico. The network he initiated soon grew to include Native Americans from the United States and Canada as well. In one case, members of the Tohono O'odham tribal council in the United States visited Mayan residents of the Sian Ka'an Biosphere Reserve in the Yucatán region of Mexico, to learn how they dealt with the pressures and opportunities of ecocultural tourism and other economic transitions. Within a year, Tohono O'odham tribal members had discussed and endorsed their tribe's participation in UNESCO meetings regarding a proposed reserve network in their homeland. They also held community and district meetings and established a tribally managed protected area adjacent to the core area of the existing US biosphere reserve at Organ Pipe Cactus National Monument. In this process O'odham leaders' confidence was buoyed by their exchanges with Mayan, Seri, Cucupa and other indigenous participants in proposed and established reserves.

Visits between communities may involve exchange of materials – such as a traditional crop variety lost by one community but maintained in another – as well as of ideas and perspectives. Keep in mind, however, that practices which work well for one community may, due to cultural differences, face resistance in another [98]. Consulting ahead of time with all parties due to take part in an exchange is one way to identify and negotiate potential stumbling blocks.

3.2.5 In situ *management and land tenure*

Work to conserve plants and other natural resources is likely to be more effective when it dovetails with adjacent concerns of local residents. For many rural peoples, few concerns are more pressing than securing their rights of ownership, also called **sovereignty** or **tenure**, over the land they farm and products they harvest. All cultures have clearly defined ways by which individuals establish and maintain ownership over land and resources. But throughout Latin America, Asia and Africa, many people face the disquieting reality that their customary tenure rules, passed on through oral traditions and mutual adherence, may no longer confer undisputed ownership. National governments are all too capable of brushing aside customary tenure relationships when miners, power companies, logging operations and ranchers covet the forests, rivers and soil of rural communities. Elsewhere, customary tenure systems may struggle to resolve internal debates over land ownership exacerbated by population growth, migration and changes in communities' natural resource base [122].

Given these realities, it is important for conservationists to assess how their actions to protect plant resources are likely to impact local tenure. The precise relationships involved in customary tenure regimes are too intricate and variable to detail in the short amount of space available here. But we can offer several general guidelines for managers seeking to learn more about local tenure arrangements in and around protected areas.

To begin with, there are three distinct social arrangements of tenure rights [36]. First are individual or family holdings, such as farmsteads, ranches, field plots, forest gardens and other forms of private ownership. Second, there are communal lands, such as forests, coastal waters, pastures, wetlands, woodlots and other areas where management and ownership are spread across a community of people. Community members cannot prevent each other from using the commons (though they do usually establish mutually agreed upon regulations to manage their use), but non-community members are generally excluded. Finally, there are public lands owned and managed by national, state and regional governments, such as national parks, forest reserves and military reservations. On these lands, governments regulate access and have the power to exclude present, past, or potential user groups.

Within these categories, tenure arrangements can take on additional complexity. For instance, within family holdings, husbands and wives can each hold authority over particular subdivisions of natural resources (Figure 3.13). In addition, there can be differences between rights to land and rights to the trees, forage grasses and other plants found on the land. Among the Ibo of Nigeria, valuable trees that germinate naturally on a private plot of land are the property of the plot owner. However, any tree that is planted by human hands is the property of the planter, regardless of who owns the land where the tree is planted. Moreover, on Ibo communal fallow land, valuable trees are the joint property of all eligible community members, while on communal land under cultivation, trees are the property of the family farming around the tree [48]. It is not hard to see how even a seemingly straightforward project like tree planting to increase local fuelwood supplies needs to take account of local tenure arrangements.

The most efficient way to sort out how conservation actions will interact with tenure arrangements is to elicit from community members the existing patterns of tenure overlap. Rural appraisal methods such as community workshops, field transects and map-making can all help to document and make sense of customary tenure arrangements. However, as Box 3.12 indicates, there are a number of subtleties to keep in mind for discussions of tenure between outside professionals and local residents [36].

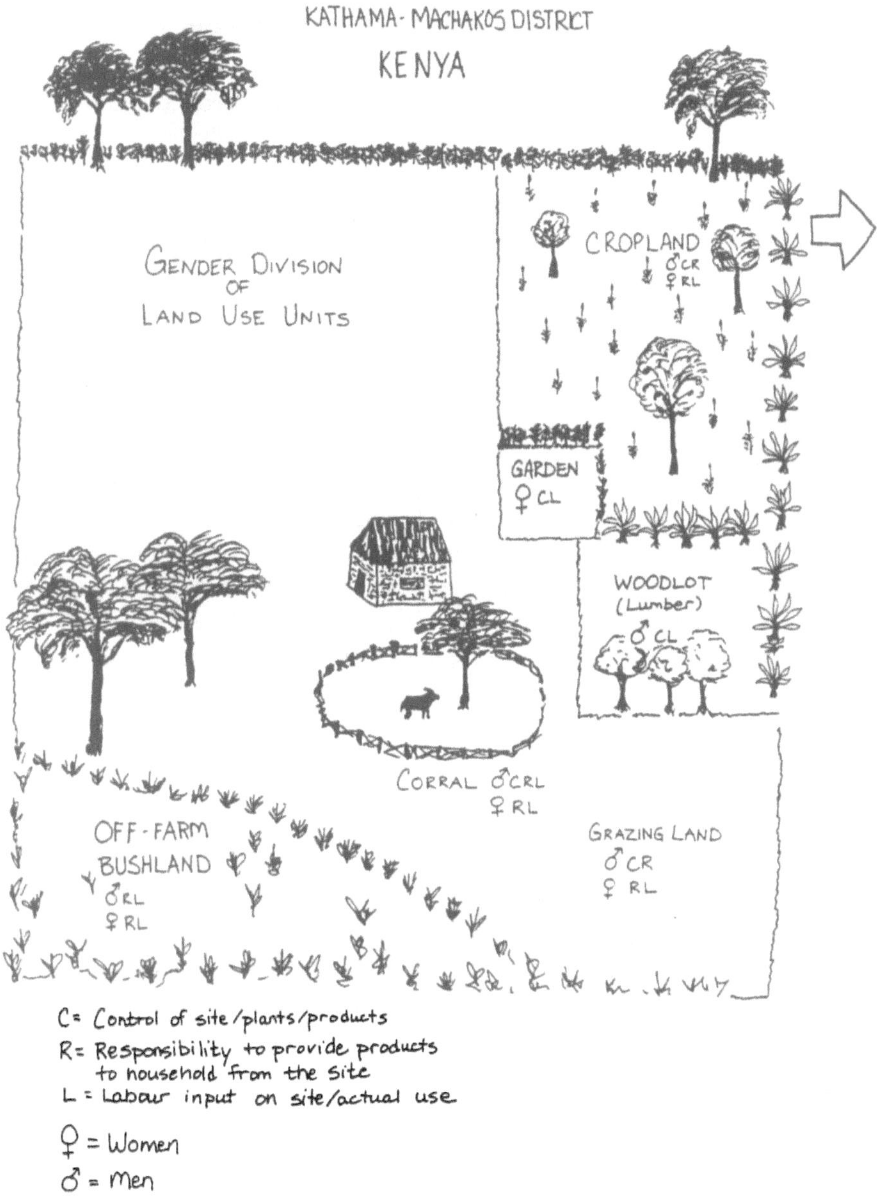

Figure 3.13 Gendered resource map from Machakos District, Kenya, showing how tenure over land and household resources can vary within a household. Note that men have control over most land, except for garden plots. However, responsibility for obtaining various products from different areas, and for contributing the labor to manage and cultivate land, are divided quite finely between men and women, creating a far more complex tenure situation than it might seem at first glance. (Reprinted by kind permission from Rocheleau *et al.*, 1995. Map drawn by M. Schmuki.)

Box 3.12 Discussing land and resource tenure

Tenure is best investigated through individual or small group interviews. Large village-wide meetings or meetings with prominent local spokespersons can be useful for identifying which individuals may be particularly knowledgeable about local tenure arrangements. However, such meetings may not yield accurate information in communities where tenure is a socially and politically charged subject, and are less likely to be an adequate forum for the views of less assertive residents, such as women or poor families. Transects and other field excursions with residents are equally valuable occasions for conducting tenure interviews.

A second important issue is how to frame questions about tenure. Direct questions such as 'Who is the owner of these trees?' or 'Can this land be sold?' carry many assumptions, and can have multiple interpretations. A more effective strategy employs indirect questions that link tenure with actions or choices, such as: 'If a newly married couple wanted to build a house, where would they get the thatch for the roof? How far away is that place? Do they need to ask anyone's permission to cut it? If so, whose permission? Why them? Can they cut thatch at any time? As much as they want? Can others take thatch from there? How can people from other villages be prevented from cutting thatch there?'. Through such questions, the nature of tenure rights will emerge gradually, interrelated with other important social, economic and cultural factors that influence local residents' decisions and actions [36].

In summary, land managers and conservation professionals can address and account for tenure issues by taking the following steps:

- recognizing the existence and validity of local tenure systems;
- talking and meeting with residents to document and understand the intricacies of tenure;
- exploring ways to manage plant resources within local tenure arrangements, so that tenure benefits resource conservation – and vice-versa.

When conservation activities are designed and implemented in ways that strengthen customary tenure, then local support for designation of biosphere reserves and similar conservation areas can be substantial. In the Indonesian province of Irian Jaya on the island of New Guinea, the Wasur National Park was established within the traditional territory of 13 villages. Conservationists Ian Craven and Wahyudi Wardoyo [55] conducted a mapping exercise as part of the park establishment process and documented that village tribespeople had undisputable claims to all 413 810 ha of the park. Villagers were initially suspicious about the park, but gradually came to favor the idea once park officials and local NGOs obtained formal, written recognition by the Indonesian central government of villagers' continued rights of residence and land use. Since the park now provides greater tenure security than that offered by Indonesian national law in general, neighboring villagers have expressed interest in becoming park residents too. 'Why are our village and land outside the park boundary?' asked one landowner at a community workshop on park management [55].

Halfway around the world in the arid northern Mexican state of Chihuahua, the Bolson de Mapimi Biosphere Reserve fits equally well with local tenure patterns. Legal authority to determine land use for 99% of the reserve rests in the hands of private landholders, the government agency for land reform and the rural credit bank. However, this biosphere reserve has gained local recognition and acceptance, in no small part because land owners have found that the presence of the reserve strengthens their land rights, especially when warding off unwanted poachers [115].

3.2.6 *Addressing local plant resource use*

Local use of plant resources and native habitats is likely to join land tenure as a prominent local issue for any *in situ* management effort. We have already illustrated how traditional rural cultures harvest plants and modify habitats for a variety of ceremonial and subsistence purposes and often manage their use in ecologically appropriate ways. When establishing or managing a protected area, a fundamental task is to determine current and historical uses of plant resources, not only by landholders but also by less visible members of the surrounding communities.

In some cases, people will travel great distances seasonally to harvest a single resource. The use of ocean shellfish to produce a textile dye known as *purpura pansa* was historically restricted to one Mexican indigenous group which lived inland, on the other side of a mountain range. Their traditional rights to the shellfish resource became imperiled by coastal development, which restricted access and eliminated shellfish habitat, and by Japanese industries interested in producing the dye. Fortunately, anthropologist Marta Turok documented the group's historic *purpura pansa* use to the degree that Mexican government agencies granted indigenous artisans exclusive rights to the resource and restricted further development in areas where the shellfish occurs. The government took these steps even though the indigenous group does not technically have tenure over land along the coastline.

Unfortunately, traditional ways of managing plants and wildlands can also give way to ecologically unsound, exploitative harvesting as communities experience population growth, greater contact with external market economies and other social changes [163, 236]. In many cases, conservationists seeking to document and strengthen traditional plant management patterns will simultaneously have to address the degradation of plant resources fueled by rising commercial or market value. A host of variables comes into play when developing strategies to make harvesting of plant resources ecologically and socially sustainable: the part of the plant harvested, the habitat preferences of the plant, regenerative capabilities of the plant, current and potential markets for the harvested products, ability of current resource users to prevent outsiders from harvesting the plant and so on. We touch upon some of these issues here, but also recommend another book in the People and Plants series, *People and Wild Plant Use* by Anthony Cunningham, for all conservation practitioners involved in projects addressing local wild plant use.

In all situations, however, managers need to be cognizant of the social and economic factors driving changes in local plant use. Resource conservation is likely to be only one of many community development priorities, such as schools, health care, wage employment, or access to manufactured consumer goods. Ethnobotanist Paul Alan Cox found this out at first hand while doing fieldwork in the Samoan village of Falealupo in the western Pacific [16]. Villagers had long protected a large communal area of lowland rainforest, from which they harvested medicinal plants and timber for canoes and household items. However, in 1988 Falealupo was ordered by the Samoan national government to build a local school. It appeared that the only possible way to raise the $65 000 needed to build the school was through commercial logging, and after protracted debate the village decided to contract with a logging company. Cox and other researchers were in Falealupo when the first bulldozers arrived and:

> villagers wept as the bulldozers pushed over the trees ... [We] shared the villagers' alarm and despair over the destruction of the forest. [We] posed a question to the village council: Could they stop the logging if [we] undertook to raise enough money abroad for the school?

The villagers' first reaction to this offer was one of suspicion, fearing it another attempt by outsiders to gain control over their forest. But after a long series of community meetings, they agreed to give Cox a chance, since they had known him for years. In the end, the community accepted the funds raised by Cox's Seacology Project for the school and cancelled the contract with the logging company. A coalition then brokered an agreement where Falealupo residents retained full ownership of the forest and rights to use forest products for home subsistence, but pledged to protect it from logging and other destructive uses for 50 years.

As this case illustrates, managers' attempts to influence local use of plant resources are most likely to succeed when they offer residents alternative means of meeting community development goals. In recent years, a number of organizations have begun exploring ways of linking the conservation of useful plants to the generation of local community development benefits. Some projects have promoted new economic activities, such as ecotourism, to generate support for protecting the habitat needed by plants. Others emphasize improving harvesting techniques, value-added processing of plant products and diversifying markets to increase the financial returns received by local plant harvesters [50]. A number of projects encourage the horticultural propagation at home or on-farm of wild-harvested species as a way to relieve pressures on remaining wild populations. This approach has been recommended and applied for medicinal plants in Africa [60], bulbs in Turkey [14] and cycads in Mexico (Box 3.13). While all of these approaches can support the *in situ* management of plant resources, the best predictor of their success is likely to be the degree to which the projects' objectives are understood (and their implementation supported and directed) by local residents as well as conservation professionals [143].

Box 3.13 Linking in situ management with local economic needs

Mexico possesses a rich cycad flora with 35 recorded native species, but these ancient cone-bearing plants face heavy pressure from habitat loss and destructive harvesting for the horticultural trade. At the community of Monte Oscuro in the state of Veracruz, the Francisco Javier Clavijero Botanic Garden has been working since the early 1990s with small-scale farmers to reduce pressures upon one heavily exploited cycad species, *Dioon edule* [238].

Botanic Garden staff have engaged residents in both *in situ* protection of remaining cycad stands on community lands and nursery production of cycad seedlings for sale to horticultural traders. At Monte Oscuro, villagers designated an 80 ha parcel of xeric thornscrub forest, left uncultivated due to shallow soils and steep slopes, as a protected reserve for the wild *D. edule* population present there, making the plants off-limits for harvesting. In addition, farmers dedicated half an acre (0.2 ha) of farmland for a community nursery to raise seedlings of *D. edule* and three species of *Chamaedorea* palms, all with high horticultural value. Farmers donated their labor to construct the nursery and received training from the Botanic Garden in essential horticultural techniques, such as compost preparation and how to harvest and germinate cycad seeds.

Seed supplies for cycads and other plants are taken from nearby wild stands under a managed field collection program. Initial population studies of *D. edule* by Botanic Garden staff revealed high seed production and seedling mortality, suggesting that a pool of surplus seeds was available for harvest. The project is now experimenting with reintroducing *D. edule* plants from two to six years old, to offset any possible downturn in seed-based recruitment due to harvesting.

Since beginning the operation in 1994, the nursery has generated approximately US$1500 in revenues for the El Palmar community from cycad sales. According to Botanic Garden representatives Andrew P. Vovides and C.G. Iglesias (238), the project has:

put 'value' on the habitat, and illegal extraction and habitat destruction has already been discouraged by the local inhabitants. There have been two instances where the local peasants have reported illegal collecting to the authorities, leading to arrests and confiscation of plants.

Since the Monte Oscuro project began in the early 1990s, nurseries have been established for four other cycad species in four additional southern Mexico communities. By linking *in situ* management with an *ex situ* revenue-generating enterprise, the project has elicited strong support in rural farming communities for protecting wild cycad populations. Local residents recognize the wild stands as essential for the nurseries' success, and are committed to their full protection.

3.2.7 Communication across cultural boundaries

Effective communication between local residents, resource managers and conservation professionals results from what is said, the context and tone in which it is said and when it is said. Frequent communication can make unlikely projects succeed and a lack of communication can lead apparently straightforward projects down a sure path to failure. A common complaint of O'odham Indians living near Organ Pipe Cactus National Monument in Arizona is that researchers and land managers do not visit them frequently enough, as if they 'don't matter', or their friendships and knowledge are not important.

Regardless of what agency, institution or organization land managers represent, they should be prepared to respond to local residents as individuals, with one-on-one ethical responsibilities. A main reason the Bolson de Mapimí biosphere reserve in Mexico has remained functional is the individual friendships between local people and researchers. Certain species have been treated well because local residents identify them with particular researchers. For instance, no one would think of harming 'Gustavo's birds', for Gustavo, an ornithologist, had endeared himself to locals [116]. Similarly, others note that villagers are less likely to steal from their neighbors' forests than from government land [164].

It is also important that conservationists concentrate on keeping their word and avoid creating, however unintentionally, a sense of disillusionment by promising more than can be achieved. In this way, trust can be built up over time through the repetition of objectives completed and mutually celebrated. Repeated visits to sites with community members also help to demonstrate conservationists' personal commitment to the local community over and above official business obligations. For example, the Bolson de Mapimí research station's sponsorship of festivities and sharing of resources such as transportation has reinforced good relationships with local communities [116].

Residents and professional land managers may theoretically agree on overall goals – such as the importance of conserving declining plant resources – but find that their interpretations and priorities for how to achieve the goals differ greatly, producing conflict. In such situations, it is important for community leaders and managers to be able to sit down, present and review each other's interpretations, and arrive at specific goals mutually understood by all parties.

Community meetings and workshops (Box 3.14) offer a format for such discussions. Societies differ in the degree to which they have developed informal institutions for discussing and debating common problems. Among communities with strong traditions of oral debate, establishing regular meetings to discuss and review issues relevant for resource conservation and land management can be an effective way to generate and sustain community collaboration. Such meetings, open to all community members who wish to attend, offer regular contact and exchange of viewpoints between managers and local residents. Meetings are also a means of spreading information about a project's progress or potential impact.

When facilitating community meetings, conservation professionals should be aware of several issues. As with interviews, not all community members will necessarily contribute equally in meetings. One way to ensure that the views of more marginalized community groups are heard is to promote smaller focus-group meetings open only to marginalized individuals. Summaries from these focus groups can then be presented in larger community meetings to stimulate debate and discussion. The style and structure of meetings are also likely to vary between cultures. For instance, the pace of debate may be slower or the style more formalized in traditional societies, compared with what conservation professionals may be accustomed. In some situations, the presence of a skilled moderator or facilitator is invaluable, both to keep discussions on track and to counterbalance individuals, such as political leaders, who may seek to usurp meetings for their own agenda.

Box 3.14 Holding a group or community meeting

Beforehand

1. Review what you would like to achieve by holding a meeting, and whom you would like to participate. Prepare a checklist of all the things you must do in order to arrange the meeting, along with a timetable for accomplishing them.
2. Contact local leaders or other relevant officials to let them know about the meeting and what it will cover.
3. Choose a meeting time and place that will be convenient for people to attend. Check with both men and women locally to confirm a good time and date.

4. Publicize the meeting well in advance – at least a week. Tell people not only the time and place, but the purpose of the meeting as well. You can announce the meeting with posters, public announcements, radio messages and direct visits to key people.
5. If you expect the meeting to last several hours, plan on having food and drinks available for participants. Often it is possible to arrange in advance with a local resident to prepare a simple group meal.
6. Plan an agenda for the meeting, and write it out on a large sheet of paper or other display board. If you expect to use visual aids at the meeting, such as handouts or a video, establish in advance how you will present them and what supplies or equipment you will need.
7. Think about how you can encourage discussion and two-way exchange at the meeting. You may want to prepare discussion questions in advance, plan for small-group discussions during the meeting, or hold a live demonstration of any new techniques or technologies that will be debated.
8. If you do not have prior experience of conducting a meeting or workshop, consider arranging for a facilitator. An ideal facilitator is someone who is skilled as well as familiar to and respected by the community, such as a schoolteacher or extension worker. Contact such individuals in advance if you think you might need help.

During

9. At the start of the meeting, introduce any unfamiliar outsiders to the audience and explain why they are attending. Do not forget to acknowledge prominent local individuals in attendance. If the meeting is small enough, consider having all participants introduce themselves.
10. Facilitating the meeting from a blackboard or large paper flip chart can allow all participants to see what the meeting will cover – and also what you are noting down as the meeting progresses. Even if participants cannot read what you write, they at least know that you are compiling a record of the meeting.
11. Keep introductory remarks short and to the point. Be sure to review clearly the purpose, goals and agenda of the meeting. Place the meeting in context as well: what past events or desired future goals are involved?
12. If you have any reason to think the meeting will involve sensitive or conflictive issues, it is best to begin the meeting with general topics likely to find broad agreement, and then work into more sensitive topics

gradually. When contentious points arise, insist that participants express their opinions respectfully and stay focused on the subject of the meeting. Try to identify potential paths that can lead the group towards resolution.

13. Try to follow the stated time schedule for the meeting. As the meeting progresses, inform participants how well they are keeping on schedule. If it is clear you will not complete all agenda items within the allotted time, postpone certain items for a subsequent meeting.
14. To finish, summarize the meeting's accomplishments, review decisions made and tasks assigned, and remind everyone of agreed-upon follow-up activities.

Follow-up

15. Review how the meeting went. Which people voiced their opinions? Who was present but silent? Which segments of the community did not attend? How did the composition of participants affect the goals and accomplishments of the meeting?
16. As soon as possible write up the minutes from the meeting, including the purpose of the meeting, major decisions reached, points agreed upon and follow-up responsibilities.

Source: [62].

Ideas and concepts can often be communicated more effectively if a variety of media are utilized, such as radio and audiovisual presentations. Radio in particular is an important communication channel by which large numbers of rural residents in Asia, Africa and Latin America receive news of the world outside their village, valley, or farm. A variety of radio show formats have been adapted to convey a conservation message, including formal documentary programs, discussion roundtables featuring local residents experimenting with a conservation innovation and plays or skits that have a conservation theme [29]. In the county of San Carlos in Costa Rica, community-oriented radio programs – hosted, directed and produced by local residents – are very popular [124]. Such programs, which focus on community issues, are a natural venue for disseminating messages with a conservation theme.

In regions where people do not have access to radio or have low literacy rates, awareness of conservation issues can be imparted effectively through locally familiar symbols, stories and communication traditions. One successful forestry education program in Indonesia employed characters and themes from popular Indonesian shadow puppet theater to convey conservation principles [164]. In

Uganda, a theater group organized to revive traditional performing arts has collaborated with soil scientists to use music, dance and theatrical performances to communicate techniques for maintaining soil fertility [199]. The success of this effort has led the theater group to tackle topics of endangered species, biodiversity conservation and traditional ecological knowledge as well.

3.3 Summary

As a summary exercise, take a look back at Table 3.1 and evaluate how the techniques and approaches described in this chapter each address the underlying chapter goal – moving conservation towards true collaborative management. The rural appraisal techniques (section 3.1) collectively represent a form of **participation by consultation.** Whether or not they move further to **functional** or **interactive** participation depends not so much on the methods themselves, but on two other factors: the degree to which local residents participate in shaping the focus of appraisal and the range of issues investigated; and the effectiveness with which all parties involved follow up on appraisal activities and actively apply the information learned to management. The process of participatory rural appraisal (section 3.2.1) is designed to put local resident–resource-professional collaborations on a track to interactive participation and, if the community resource management plan is successfully implemented independently, to achieving **self-mobilization.** Note, however, that PRA can easily be limited to functional participation if sufficient responsibility or control over the process is not vested among local participants.

The common practice of using local residents as assistants for scientific research usually does not move beyond **participation for material incentives.** The process of participatory action research (section 3.2.2) is designed to make research a form of interactive participation. Further, as the example of World Neighbors in Bolivia indicates (Box 3.11), the skills that local participants can gain through PAR-related seminars and workshops can promote self-mobilization among community members. However, again a word of caution is required: if participatory action research is not implemented with a careful eye to involving local residents in all stages, it may not move beyond functional participation.

A similar pattern applies to paraprofessional training (section 3.2.3), which transmits skills and experience to local residents that can be self-mobilizing. However, if self-mobilization is not an active goal of paraprofessional training, such training can be limited to participation for material incentives, and even **passive participation** if the training is implemented particularly poorly. Community exchanges (section 3.2.4) likewise are best viewed as a tool towards self-mobilization, but this is a realistic benefit only when applied within programs that take goals of self-initiation and local decision-making to heart.

The final themes of this chapter fit relatively loosely within those outlined in Table 3.1. Land and resource tenure (section 3.2.5) and local plant use (section3.2.6) are best seen as issues with potential to make participation interactive, by virtue of their being extremely important within the lives of many local residents. However, they are all too easily addressed in a passive fashion in many cases. Likewise, careful consideration of **how** to communicate ideas and messages (section 3.2.7) is likely to help interactive participation succeed, but will not automatically lead to it.

We suggest that achieving truly participatory management of plants and other natural resources depends closely on the perspective and approach of the individuals involved in resource management; it is not something that flows automatically from a given set of tools or techniques. On a fundamental level, collaborative management of plant resources must express a mutual recognition of the value of local knowledge and experience – as well as reciprocal acknowledgement of the unique contributions that resource managers and other professionals can make. These themes will resurface continuously throughout the next three chapters as we explore ways to identify conservation priorities, design *in situ* management plans for plant resources, monitor and evaluate management activities and address the unique issues involved in conserving plant resources in traditional agricultural landscapes.

4

Setting priorities and planning for management

Figure 4.1 Roadside stand near Alamos, Sonora, Mexico, selling products made from wild chile peppers (*Capsicum annuum*). Pickled chile fruits are displayed as well as live plants suitable for transferring to home gardens. Harvesting wild chiles for the spice trade and for home use is an important cottage industry in rural areas of northern Mexico. Any effort to conserve wild chiles as a genetic resource must take into account their importance as an ethnobotanical resource in rural economies. (Photograph: G.P. Nabhan.)

An important part of any successful conservation effort is the ability of managers and field workers to use limited resources – funds, personnel, equipment and knowledge – to their greatest potential. Conservation teams need to prioritize where to put their time and resources, particularly at the onset of projects. This chapter presents a process to help you identify and set priorities for managing plants and other natural resources.

In some cases, setting priorities may commence with gathering basic ecological information on the plants that occur in or around a protected area. It may also involve gathering baseline social data with local residents who depend directly on plant resources for their livelihood, such as craft artisans. Thus we begin this chapter by reviewing approaches to inventorying useful plants and compiling the information that inventories provide. When properly applied, systematic inventory methods can generate a wealth of information for prioritizing plant conservation in a region.

Setting priorities also requires a way to compare useful plants, their habitats and co-occurring species, according to characteristics important for conserving plants, such as their rarity or their value within local cultures. This chapter lays out a format for such comparisons. It includes a review of several ecological criteria that make species, populations and habitats priorities for conservation, and a discussion of how to characterize threats to plant resources. Finally, identifying priorities will be of little practical benefit unless they are translated into concrete targets and actions for management. We conclude with a step-by-step approach for developing an adaptive management plan targeted at the conservation priorities that you identify.

4.1 Identifying what is out there: inventories of useful plant resources

To obtain the information necessary for managing natural resources and protected areas, conservationists typically turn to **inventories**. Broadly defined, an inventory is any quantifiable, repeatable method of sampling, classifying and measuring an environmental feature or cultural pattern. Among the natural and cultural resources commonly inventoried by land managers are timber stands, archeological sites, soil types and vegetation associations, to name but a few. Inventories are possible on many scales – from an entire watershed or mountain range down to a small forest fragment, from 1000 km^2 of wildlands to 10 m^2 of backyard garden.

There are at least five ways to inventory plants in native habitats:

- **species checklists** that compile and geographically annotate species occurrences;
- **biotic community surveys** that map the distribution of habitat types and vegetation associations;
- **ethnobotanical inventories** that document local knowledge and use of plants;

- **ecogeographic surveys** that identify and map areas with high concentrations of wild relatives of crops, medicinal herbs, or other plant resources;
- **population genetic studies** that quantify the amount and distribution of genetic variation within a species and its populations.

The first two are general or **baseline** inventory approaches that provide valuable background material. They can reveal target resources and sites for plant conservation, but can be a big job to do comprehensively, particularly for large protected areas or in habitats that are difficult to sample, such as mature tropical forests. The last three inventory methods document specific dimensions of ethnobotanical and phytogenetic plant resources that can easily go undescribed in more general inventories. Managers may wish to employ them in situations where specific information about useful plants is desired in greater detail. Ethnobotanical inventories are particularly important because local residents' priorities for conserving plant resources will not always be the same as those of conservation professionals, and ethnobotanical inventories can reveal those local priorities. Our discussion of these five inventory techniques will remain relatively general. In addition to the references listed in the text, Appendix 1 lists literature to which managers and conservation professionals can turn for more detailed information on these methods.

4.1.1 Inventories of species richness

Compiling species occurrences in a protected area or region is the most common type of baseline plant inventory. Ideally, such inventories go beyond merely listing species and include an initial assessment of other components of plant diversity, such as patterns of species distribution and abundance. This can be accomplished by charting species occurrences on maps; coding their occurrences relative to landscape units identified by topography, soils and other physical features; and making rough estimates of abundance or cover (discussed in section 4.1.2). Although such inventories may only give an initial impression of species presence or absence, called **alpha diversity** by ecologists, they are the first steps towards a full assessment of plant diversity.

When surveying a region or area for plant species diversity, aim to make your sampling as representative of the entire area as possible. This means visiting the different habitats that may be evident within a protected area; covering a range of different elevations, slopes and **aspects** (the compass direction a slope faces – influencing, for example, how much sun it receives); and sampling vegetation of various successional ages (section 5.4.4). As discussed already (section 2.1.2), long-time local residents who have a deep familiarity with the diversity of sites or habitats in an area can often identify appropriate areas to sample and help to make a survey more representative. Also be aware that for protected areas that are very large, completing a plant species inventory that is truly representative is likely to be a long-term project. This is particularly true for mature tropical forest habitats that tend to have high numbers of species and be dominated by tall, irregularly flowering canopy trees and lianas that are difficult to collect in fertile condition.

The physical documentation involved in a plant species inventory consists of collecting **voucher specimens** for each suspected plant species encountered during an inventory. If managers or conservationists are contemplating such documentation, we recommend consulting with an experienced botanist familiar with collection techniques. The first manual in the People and Plants series, *Ethnobotany* by Gary Martin [133], also provides a detailed overview of how to collect and preserve high-quality botanical specimens.

To make vouchers, specimens gathered from the field are placed between sheets of newsprint, which are then bound together with blotter papers and cardboard or aluminum corrugates in a plant press (Figure 4.2). The press flattens the

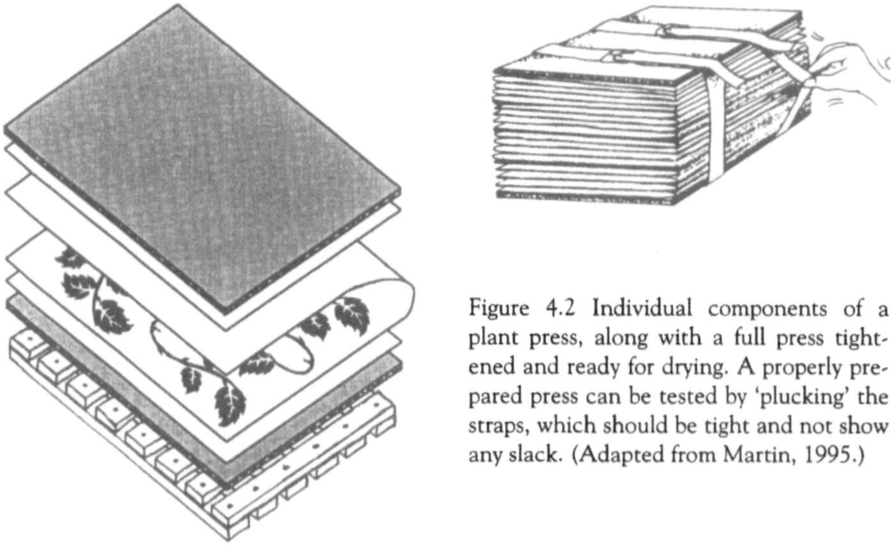

Figure 4.2 Individual components of a plant press, along with a full press tightened and ready for drying. A properly prepared press can be tested by 'plucking' the straps, which should be tight and not show any slack. (Adapted from Martin, 1995.)

specimens and facilitates their drying, which is usually done by placing the press over a heat source or in front of blowers. Once dried, vouchers are usually deposited at a herbarium for mounting, identification and storage. Pressing and documenting quality botanical vouchers is a skill unto itself; several guidelines for collecting are given in Box 4.1.

All specimens must be labelled with a unique identification number, composed of the name of the person who makes the collection and a number (for instance John Tuxill #322). Most botanists assign their collection numbers sequentially over the course of their lives; but some include the year in each collection number and begin a new number sequence annually.

Box 4.1 Collecting plant specimens

Since reproductive parts are often required for botanists to confirm a plant's identity, each voucher should include flowers or fruits whenever possible, in addition to vegetative parts of the plant. Large fruits are easier to view and dry if they are sectioned longitudinally or cross-wise into sections no more than 2–3 cm thick. This exposes ovary divisions, seeds and other anatomical features useful for identifying the specimen. Since flower and fruit colors often fade or change as a specimen dries, it is important always to note the colors of the fresh specimen before drying.

Specimens should fit the dimensions of a standard herbarium sheet (approximately 28 × 42 cm). Herbs, epiphytes and other small plants can be placed into a press whole, bending or trimming the stem where necessary to make the plant fit on a page. Cuttings from trees, shrubs and large herbs should show the plant's branching pattern. Identification can also be made easier by flipping over at least one leaf of the specimen, so that both front and back leaf surfaces are revealed in one glance.

Palms, similar large-leaved plants and succulents like cacti can all be daunting to collect or preserve. For these plants, it is essential to measure and describe the dimensions of the leaves, stem and other plant parts too large or thick to fit in the press whole. With palms, experts recommend counting the number of pinnae (leaflets) on an average leaf, measuring its length and width, and then pressing the tip, basal pinnae, and mid-section of a leaf. Inflorescences and fruit can be pressed on a separate sheet – but make sure all are clearly labelled with identification (ID) numbers to correspond with the leaf specimens from the same collection. For succulents, it is best to measure and record the stem dimensions, and press cross-sections and longitudinal stem sections [133].

When collecting plants, you should record descriptive information for every voucher. Most botanists write down this information in notebooks when a specimen is placed in the press, and then transfer it to labels that are included with each voucher. These labels are eventually mounted on the herbarium sheet with the specimen. In addition to the collection number, standard information to note for every voucher includes [133]:

- **scientific name** (added afterwards, if not known by the collector);
- **location** of the collection, described as completely as possible, indicating the country, state or province, district and distance and direction from the nearest town; also include approximate longitude, latitude and altitude if available from maps or from a global positioning system (section 3.1.3);

- **appearance** of the plant, including color of flowers, fruit and other distinguishing parts, the plant's size and growth form, bark characteristics and presence of resins or other exudates;
- **relative abundance** of the plant – common, rare, occurring in thick stands or as scattered individuals;
- **description of the site** where the plant was found, including **slope** (flat bottomland or steep hillside), **aspect** (whether the site faces north, south, east or west), **soil characteristics** and **co-occurring vegetation** (particularly any dominant species);
- a list of **associated collections** of seeds or fruits for germplasm storage, bark or wood samples, whole fruits or flowers preserved in alcohol, or other items related to the plant;
- **names** of yourself and other people making the collection;
- **date** of collection.

Additional information recorded for a specimen may vary according to why a plant is being collected. For instance, vouchers collected in ethnobotanical inventories (Figure 4.3) should note local nomenclature and plant uses. All of

Scientific name: *Proboscidea parviflora* (Woot.) Woot. and Standl.

Family: Martyniaceae

Local names and languages: 'ihug (Tohono O'odham, Kokolodi dialect) devil's claw (English).

Description: Cultivated annual shrub, leaves viscid, flowers reddish-white.

Location: U.S.A., Arizona, Pima County, Tohono O'odham Reservation, Sells District, Sells Village, dooryard garden of Enos Francisco, junior. Approx. lat. 31°52′N, long. 111°51′W.

Associated vegetation: Floodwater garden surrounded by Arizona upland desert scrub; associated cultivated plants included tepary beans (*Phaseolus acutifolius*) and squash (*Cucurbita* spp.) ; wild amaranths (*Amaranthus* spp.) dominant weeds.

Biotic associates: *Perdita hurdi* bee collected cutting into corolla; *Hyles lineata* larvae collected on leaves (ID's by Floyd Wemer, deposited at Univ. of Arizona).

Uses: Dried appendages of fruit capsules cultivated as a basketry fiber for commercial craft industry; white seeds (as opposed to wild black seeds) eaten. Planted in hills 3 m. apart, 60 hills in a 90 m. by 10 m. area.

Collector(s): P.K. Bretting (PKB 79–214) and G.P. Nabhan (GPN 79–344).

Collection Date: 12 September 1979.

Notes: Germplasm deposited with Native Seeds/SEARCH, Tucson, Arizona.

Figure 4.3 Sample herbarium label from a specimen collected in 1979. Most herbarium specimens are not described with the richness of detail present on this label. Note, however, that even this specimen could be better documented, as it is missing an altitude reading.

these data greatly enrich a specimen's scientific value and aid subsequent researchers who may wish to include the voucher in ecogeographic surveys (section 4.1.3) or other kinds of studies.

Natural resource managers may often choose to implement plant inventories in collaboration with researchers from a botanical garden, university herbarium, natural history museum, or similar facility. These institutions have the expertise to collect, process, identify and curate botanical specimens, and can greatly expedite a species inventory. However, there are also benefits to establishing a community-owned herbarium or a local reference collection of plant specimens for a protected area. A community herbarium can be a valuable educational and cultural resource, particularly when focused on plants traditionally used as ethnobotanical products within the community. Local reference collections can be very useful for researchers and others who are working in an area and need to identify the plants they encounter. In some cases, local herbaria undoubtedly warrant the necessary investment in collecting supplies, plant presses, specimen dryers, herbarium cabinets and other materials. It may be possible to establish cooperative agreements with botanical institutions where they conduct inventory work, but also return a set of voucher specimens for local storage and reference. These options are always worth exploring.

Finally, if you are undertaking a comprehensive inventory of plant species richness, additional information can help you to infer what other species might have been present historically and even prehistorically. Information describing changes in plant occurrences over time offers an important window on how plant communities respond to variations in environmental conditions, human land use practices and other influences. There are a number of information sources that can help to piece together the historic and prehistoric presence of plants within an area:

- botanical specimens in regional and national herbaria – often affiliated with universities or museums;
- early accounts made by naturalists and surveyors who visited a region;
- elders in local communities who remember observing or collecting species in their youth that can no longer be found nearby, or recall first encountering plants suspected of being recent introductions;
- identification of plants in historical photographs;
- studies of the middens left by pack rats or other rodents that collect stem, seed and leaf fragments;
- studies of plant macrofossil and pollen samples from animal dung or debris deposits in dry caves or rock shelters;
- studies of pollen deposited in lake bottoms and other sedimentary strata.

4.1.2 Habitat inventories

Habitat inventories document the **vegetation associations** or **biotic communities** found in a region. The issue of how to identify community types – or whether discrete plant and animal communities truly exist – has been widely debated by ecologists for much of this century. On at least a qualitative basis, there is a general consensus that differences between vegetation associations can be discerned on a mappable scale.

When such a qualitative or descriptive inventory is done, biotic community assemblages or vegetation associations are usually characterized by structural features (such as tall closed-canopy forest vs. open savanna woodland or arid thornscrub vs. desert grassland), environmental features (such as association with a particular soil type or surface hydrology), topographic features (*terra firme* upland forest vs. *varzea* floodplain forest) and/or the presence of one or several dominant species (dipterocarp forest, pine–oak woodland). Often, biotic communities are inventoried by mapping vegetation and censusing birds, mammals and other animals along a perceived gradient (such as elevation, slope, silt to sand, riverside to floodplain edge, etc.). Another approach is to map evident boundaries between major vegetation types, such as where upland hill forest gives way to a low-lying patch of swamp forest, or mesic gallery forest hugging a rivercourse is bordered by xeric thornscrub.

Increasingly, such mapping is being done with remote sensing imagery, such as aerial photos taken at low altitudes or even high-quality satellite imagery. However, qualitative vegetation inventories can also be done on the ground as participatory mapping exercises (section 3.1.2) involving resource management teams and local residents. Forester Mark Poffenberger and colleagues [175] have experience with this approach in India:

> A proven strategy is to involve one or two members of the community as resource informants in the team from the outset to assist in making observations and in recording data from transects and quadrats. Key informant interviews with special [resource] user groups such as basket-makers or mat-weavers ... will also provide insights into collection and utilization practices. Finally, group discussions with women, artisans, farmers and migrant graziers would also benefit the team by giving them access to specialized knowledge.

Since local residents usually have their own terms and classifications for habitat types, mapping culturally defined units may also be useful. These **folk units** are often based on dominant species, uses and successional history.

Initial vegetation mapping can be combined with visual field censuses to give a preliminary assessment of abundance and composition for important species within major vegetation types. Botanists working with the Nature Conservancy (a US-based organization) have described a rapid **visual assessment** method of quickly tallying major differences in species abundances at a site (Box 4.2).

In addition to these qualitative approaches, plant ecologists and foresters have developed a variety of quantitative methods to inventory and assess vegetation.

Box 4.2 Visual assessment of vegetation composition

Which plant is more abundant in that abandoned farm plot: the evenly spaced *Cecropia* trees arching overhead or the scattered clumps of *Heliconia*? Unfortunately, in a quick glance it is difficult to make accurate comparisons of small differences between numbers of different objects. However, most people can, with consistent accuracy, distinguish differences in numbers of objects on an exponential scale, such as between one and 10 individuals, 10 and 100, or 100 and 1000, and so on. Estimates of species abundance made in this visual fashion are of limited use, since only prominent differences between populations can be evaluated, and the results are not suitable for statistical analysis. Nevertheless, visual estimates can provide an initial rank measure of abundance for different plant species. Where managers have limited resources or time to conduct habitat inventories, such information can be a useful addition to vegetation mapping or to a list of species found in a given habitat.

Source: The Nature Conservancy

Because quantitative methods involve field sampling that tends to be time-consuming and laborious, you are probably better off to pick and choose when and where to use quantitative methods, employing them to monitor or gather baseline information on resource management priorities identified by local residents and management teams, rather than involving them at the start of a broad-based inventory. Accordingly, these quantitative methods are discussed later, in Chapter 5.

4.1.3 Inventories of ethnobotanical resources

The science of ethnobotany is focused on understanding how plant resources are used, classified, perceived and managed by different communities and cultures. Thus it is not surprising that this discipline offers a wealth of methods appropriate for inventorying useful plants. Ethnobotanical inventories are as much social as they are ecological: ethnobotanists inventory plant resources by talking and working directly with people who use plants and typically know many details of the plants' ecology and natural history. An ethnobotanical inventory is essentially an opportunity for local residents to relay which plants in a flora **they** may be most interested in conserving – without this step their voice may go unheard.

This discussion here presents a broad overview of ethnobotanical inventories and there is much to the discipline, including the specifics of collecting ethnobotanical data, that we do not have the space to present. For more information on the theory and practice of ethnobotany, we recommend *Ethnobotany* by Gary Martin [133], the first manual in the People and Plants series.

In a standard ethnobotanical inventory, ethnobotanists accompany knowledgeable consultants from local communities into the field; collect samples of any plants which consultants indicate as useful or otherwise prominent; record local plant names; record which parts of each plant are used, how they are prepared and for what purposes; and make voucher collections for taxonomic identification at a herbarium. In this way, ethnobotanists compile lists of the plants used by different cultures or communities and identify particular uses for species. Ethnobotanists may also identify which local species are collected exclusively from wildlands and which are informally protected, managed, semi-cultivated, under incipient domestication, or fully domesticated. These studies provide a snapshot of the diversity of plant species named, recognized and used within particular communities or cultural traditions.

Once local experts on plant resources become acquainted with the interests of ethnobotanists and land managers, they often volunteer information about the value or history of a certain stand of trees, interactions between species, or other topics not typically captured in initial ethnobotanical interviews. It is important to remember that the information recorded on the first set of vouchers is only a starting point and is seldom exhaustive. For instance, women in a community are usually familiar with many plants about which men know few details, and field surveys that do not consult women risk overlooking a rich subset of local ethnobotanical knowledge. Managers should recognize initial ethnobotanical surveys to be a beginning and keep their doors open for additional discussions as conservation activities progress.

Some ethnobotanists have combined standard ethnobotanical inventories with ecological sampling methods to answer questions such as which habitats in a region hold the greatest concentrations of ethnobotanical resources, or where the largest quantities of plant products used by local communities are located. Quantitative ethnobotanical approaches have involved collecting samples of all plant species (not just the useful ones) within a given plot or transect in a habitat (in tropical forests typically 0.1 ha, or 1000 m^2) and then asking consultants to indicate all the species used locally. With this inventory approach, researchers can calculate usefulness percentages for different habitat types and quantitatively compare the ethnobotanical importance of different habitats for local cultures (Box 4.3).

Box 4.3 Quantitative ethnobotany in Amazonian Peru

One quantitative ethnobotanical approach has been pioneered by Oliver Phillips and colleagues [169] to study use of plant resources by ribereno communities in the Tambopata region of Amazonian Peru. Phillips's team developed an indexing system that calculates **use values** for plants, based on the average number of uses local consultants identify each time they encounter and discuss a particular plant species with the interviewer.

They applied this system in interviews with 29 Tambopata residents (men, women and youths) and calculated use values for 496 tree and liana species. The researchers then quantified the relative abundance of these useful species by tallying their abundance, based on numbers of stems, in 20 m × 50 m (= 1000 m^2) rectangular plots (quadrat sampling: see Chapter 5) distributed among six local habitat types of upland, floodplain and swamp forest, of different ages. This enabled the researchers to compare quantitatively the value that Tambopatans place on different habitats for supplying plant resources.

Their results, part of which are presented in Figure 4.4, show that mature floodplain forests are the most valuable habitats for local residents because of the abundance and diversity of useful plants in these forests. This study demonstrates how inventory results can yield important conservation and management recommendations. Phillips's ethnobotanical inventories suggest that floodplain forests should be priority habitats for conservation efforts in western Amazonia, from the point of view of local residents. Mature floodplain habitats also deserve priority because they show the greatest deforestation rates and resource exploitation pressures in

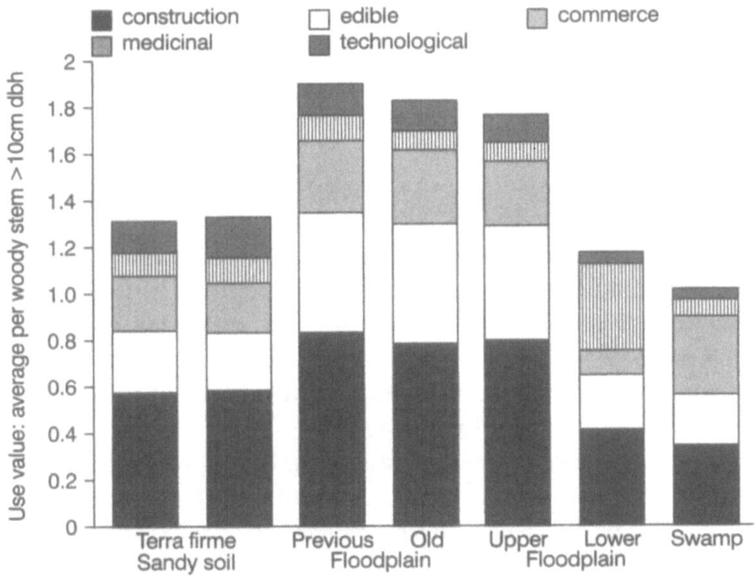

Figure 4.4 Use value compared for seven forest types at Tambopata, Peru. The Upper Floodplain, Old Floodplain and Previous Floodplain categories are all variations of mature forest. (From Phillips *et al.*, 1994, reprinted by kind permission of Blackwell Science, Inc.)

the region. Ethnobotanical studies – particularly when undertaken with a quantitative approach – can offer valuable information for resource managers and local residents alike when collaborating on conservation plans for lands and habitats that have been under traditional management and used by local residents.

4.1.4 Ecogeographic surveys

Sometimes plant resources can be inventoried most rapidly by identifying and mapping habitats or geographic regions that host exceptional concentrations of ethnobotanical resources, rare species, or other plants of interest. This type of assessment is referred to as an **ecogeographic survey**. One ecogeographic approach is to identify areas or regions that hold concentrations of different species within an economically important genus. Ecogeographic surveys are particularly useful for phytogenetic resources like wild crop relatives, since ethnobotanical inventories may not reveal their presence if such plants do not have recognized uses in local cultures.

We can illustrate the utility of ecogeographic surveys with an example from the northern Sierra Madre Occidental of Mexico [144]. This 180 000 km^2 montane region (Figure 4.5) is well known as a major center of plant diversity, home to over 4000 vascular plant species [15, 76, 200]. Until recent studies were accomplished, however, it was unclear which parts of the northern Sierra Madre held the greatest concentrations of wild crop relative species. As a result, no one knew where *in situ* reserves should be located to conserve the greatest number of species valuable for this purpose.

The first step in proposing sites for protection in the northern Sierra Madre involved mapping the locations of all herbarium records and germplasm collections for each species known to be a wild relative of a major crop. This mapping was done for common bean (*Phaseolus*) species from each of eight subregions in the northern Sierra Madre (Figure 4.6, Table 4.1) to demonstrate where collections had been made and thus documenting a species occurrence. Location information was obtained from the **passport data** deposited with each herbarium voucher and germplasm sample (Box 4.4).

The second step involved comparing the number of *Phaseolus* germplasm collections made for *ex situ* storage with the number of wild bean species recorded from each subregion by past botanical surveys (Figure 4.7). This step identified the subregions in which wild beans appeared to be undercollected for *ex situ* conservation relative to their numbers of species indicated by herbarium specimens. It demonstrated that three of the southernmost subregions (E, H and G) have the greatest richness of bean species, and that northern Sinaloa has not been well represented in *ex situ* collecting but is nevertheless very diverse. The

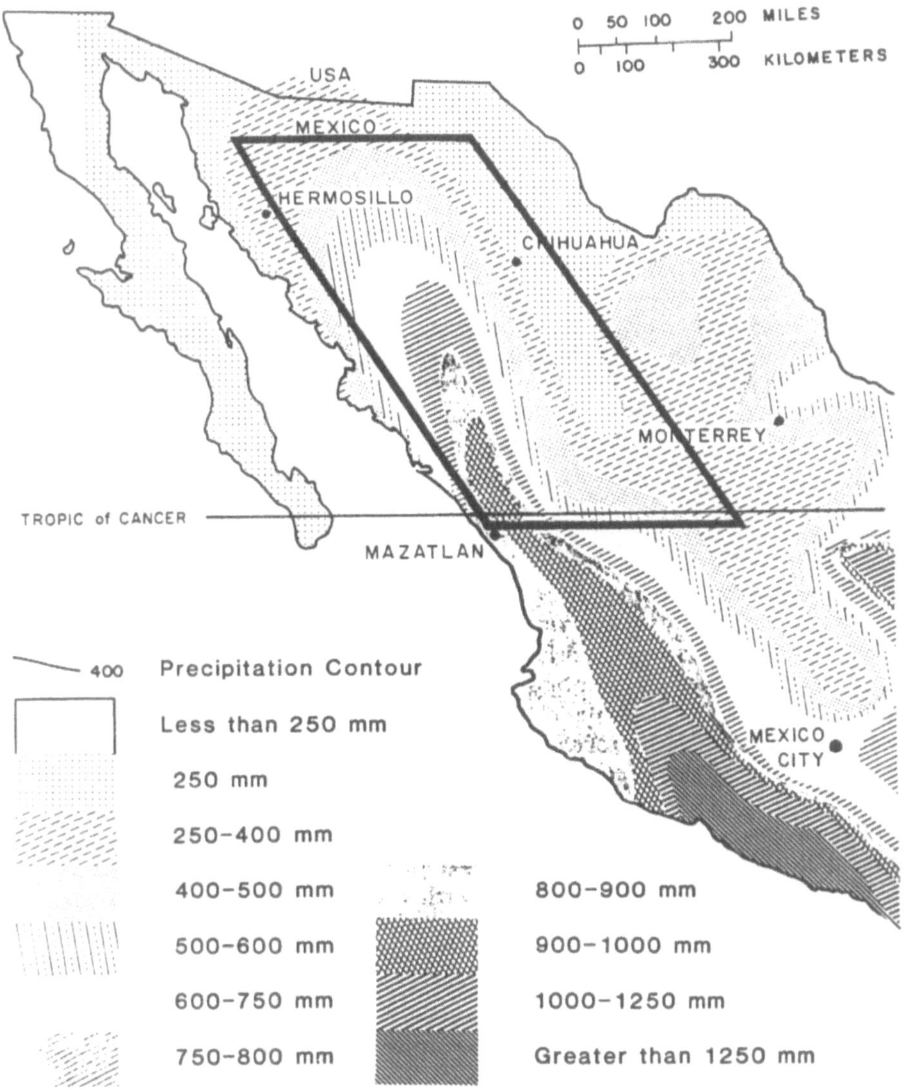

Figure 4.5 Map of northwest Mexico showing general area of *Phaseolus* ecogeography study, superimposed on a map of rainfall patterns for the region. (Reproduced from Nabhan, 1990a by kind permission of International Plant Genetic Resources Institute.)

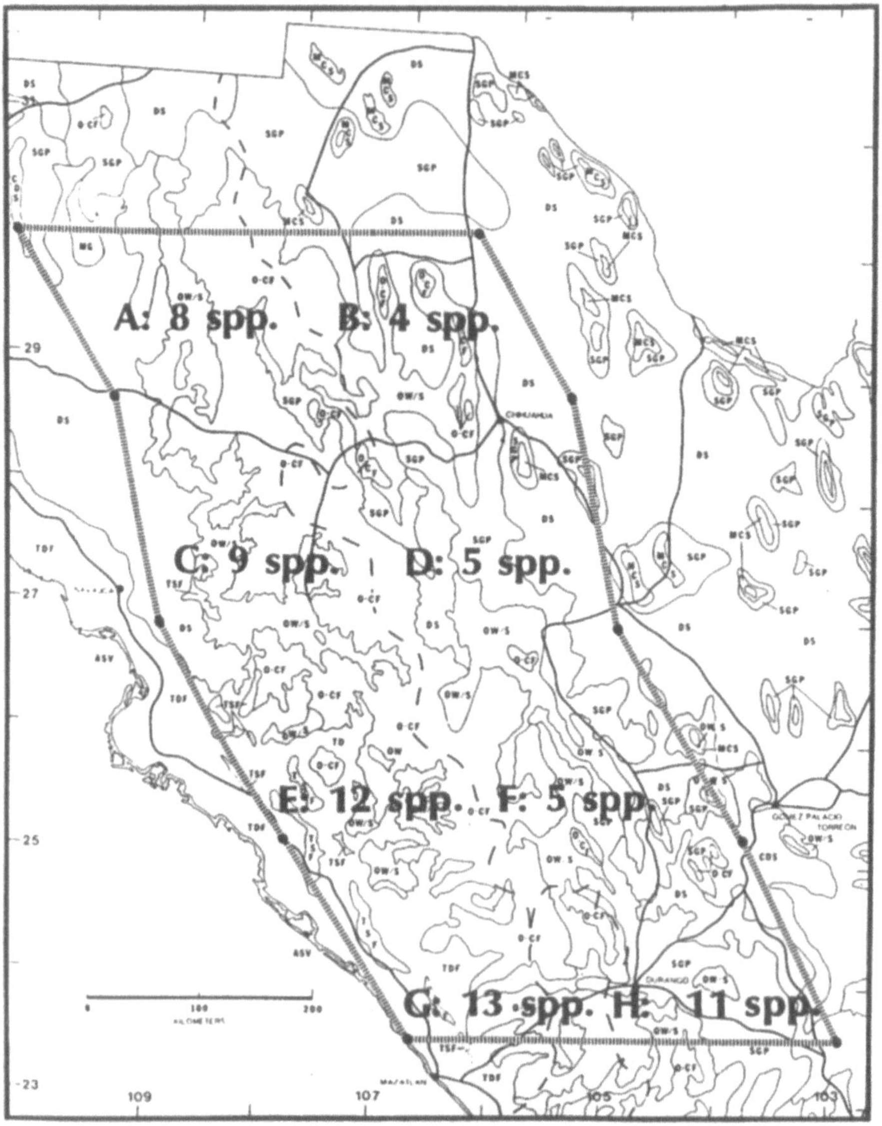

Figure 4.6 The study area in northwest Mexico (Figure 4.5) broken down into eight subregions, A–H, and indicating the total wild *Phaseolus* species known from each subregion. The study area is superimposed over a map of regional vegetation types: ASV, aquatic and semi-aquatic vegetation; CDS, columnar cacti-dominated scrub; DS, microphyllous desert scrub; MCS, mediterranean chaparral scrub; MG, mesquite grassland; OCF, oak–coniferous woodland; OW/S, oak woodland/savanna; SGP, short grass prairie; TDF, tropical deciduous forest; TSF, tropical semi-deciduous forest. (Reproduced from Nabhan, 1990a by kind permission of IPGRI.)

Table 4.1 Distribution of total *Phaseolus* collections by subregions in the Sierra Madre Occidental, Mexico (adapted from Nabhan, 1990a by kind permission of IPGRI.)

Phaseolus species	Subregions							
	A	B	C	D	E	F	G	H
P. *acutifolius*	9(1)	–	14(7)	1(–)	5(–)	4(–)	10(5)	15(3)
P. *amblyosepalus*	–	–	–	–	–	–	1(–)	–
P. *angustissimus*	2(–)	–	–	–	–	–	–	–
P. *coccineus*	–	4(1)	–	3(–)	2(–)	–	6(2)	2(–)
P. *filiformis*	–	–	–	–	–	–	–	1(–)
P. *leptostachyus*	1(–)	1(–)	10(4)	1(–)	6(1)	2(2)	5(1)	4(–)
P. *lunatus*	–	–	–	2(–)	2(–)	–	3(–)	–
P. *maculatus*	3(–)	3(1)	1(–)	1(–)	3(3)	1(–)	–	2(–)
P. *microcarpus*	–	–	–	–	–	–	4(2)	2(1)
P. *parvulus*	6(1)	–	5(2)	–	3(2)	4(2)	3(–)	8(5)
P. *pauciflorus*	3(–)	–	7(–)	–	3(–)	–	8(–)	5(–)
P. *pedicellatus*	8(–)	5(–)	8(3)	1(–)	2(–)	2(–)	3(–)	8(5)
P. *pluriflorus*	–	–	–	–	1(–)	–	–	3(1)
P. *ritensis*	6(1)	–	8(2)	–	10(6)	–	–	–
P. *salicifolius*	–	–	3(–)	–	4(–)	–	8(2)	–
P. *sempervirens*	–	–	–	–	–	–	4(2)	3(2)
P. *vulgaris*	–	–	1(1)	–	3(1)	–	1(–)	2(2)
P. *xantotrichus*	–	–	–	–	–	–	1(–)	–
Total accessions	38(3)	13(2)	57(19)	9(0)	44(13)	13(4)	57(14)	47(14)
Total species	8(3)	4(2)	9(6)	5(0)	12(5)	5(2)	13(6)	11(6)

Numbers in parentheses indicate how many of the collections were **germplasm** collections – as opposed to **herbarium** collections, where any seeds gathered were not destined for seedbank storage.

diversity of habitats, rugged topographic relief and relative inaccessibility of northern Sinaloa make it an ideal location for *in situ* protection of these wild crop relatives, a conclusion supported by these ecogeographic studies and independently confirmed by economic botanists [144].

4.1.5 Population genetic approaches

Species of known economic importance, ecological prominence, or threatened status may also warrant inventories of their **intraspecific genetic variability**, or the variation in gene pools between a species' individual populations. Such work can pinpoint populations holding rare alleles and identify important genetic factors (such as the degree of inbreeding) that operate primarily at the population level. These features can be important to the overall health or **fitness** of a species and its value as a phytogenetic resource. For instance, most of the genetic diversity within a particular species may be found in vigorous populations at the center of the species' range [141]. However, outlier or disjunct populations may be

Box 4.4 Using passport information in an ecogeographic survey

There are two ways to collect the data needed for an ecogeographic survey. One is to visit the region under study, systematically survey different habitats, and make field collections of targeted species or varieties. These might include several species that share a genus with a domesticated crop plant, or distinct varieties of an economically valuable species.

Ecogeographic information can also be obtained by reviewing previously collected herbarium specimens and gene bank vouchers for their associated documentation, or **passport data**, that describes the sites at which they were collected. For ecogeographic analysis, the key passport data include:

- place name of the collection site;
- approximate longitude, latitude, and elevation of the site;
- descriptions of the habitat in which the plant was found.

Ambiguous or vague passport data should always be interpreted carefully; for instance, one place name in a mountainous region may actually encompass a range of elevations and habitats. For hard-to-identify species, it may also be necessary to ask an expert botanist to review specimens for taxonomic accuracy. When carefully assembled, the passport information from previously collected specimens can be a valuable information resource [135].

more likely to hold rare or unique genes, particularly those which confer adaptation to environmental stresses. This means it is also important to protect populations at the edge of a species' range.

Most genetic variability analyses involve electrophoretic or DNA studies of germplasm samples collected from different populations of a species. Because such studies require laboratory facilities and can be relatively expensive, they are usually done in collaboration with university research facilities or government-sponsored research programs in agriculture or forestry. The majority of genetic variability studies of plants have been done on domesticated crops, landrace–wild crop relative gene complexes, timber trees and critically endangered species. One case where patterns of genetic variability within species are being used to guide conservation efforts is a cooperative research project known as the Central America and Mexico Coniferous Resources Cooperative, or CAMCORE [68, 69]. This project has focused upon collecting seeds from highly threatened populations of Mesoamerican conifers for *ex situ* preservation (Box 4.5). Population genetic studies focused on crop landraces and landrace–wild crop relative genepools are discussed further in Chapter 6.

The various inventory methods reviewed here – species lists, vegetation mapping, ethnobotanical surveys, ecogeographic surveys and population genetic studies –

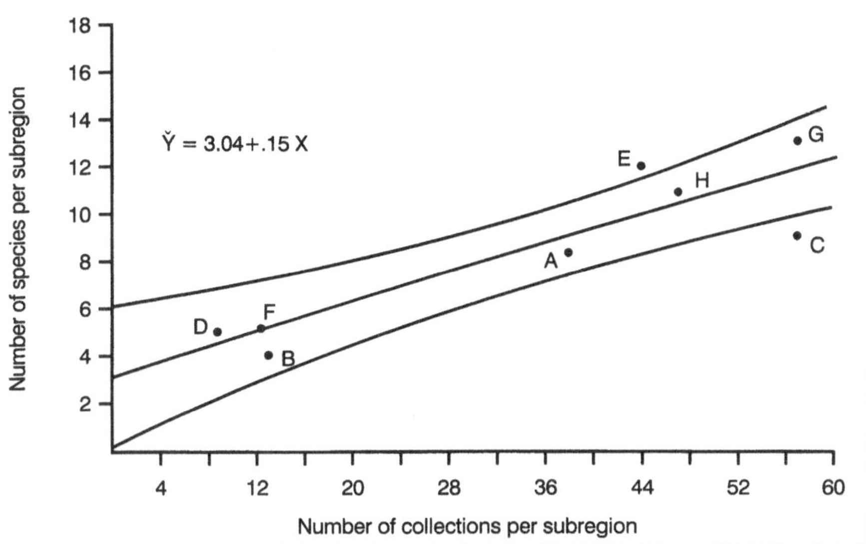

Figure 4.7 Number of *Phaseolus* species for each subregion (A–H) (Figure 4.6) compared with total number of collections made there, expressed by constructing a regression line bounded on each side by a line indicating the 95% statistical confidence interval for the comparison. Subregion E falls above the confidence interval, indicating that it is relatively under-collected, given the number of *Phaseolus* species that occur there. Subregion C plots below the confidence interval, suggesting that it has been relatively over-collected, given its species total. (Reprinted from Nabhan, 1990a by kind permission of IPGRI.)

Box 4.5 Conserving Mesoamerican conifer gene pools

Mesoamerica is a global center of diversity for pines and other conifers. Mexico alone can claim over 70 documented species and varieties of pines, more than twice the total of the continental United States (a much larger area). Recognition of the genetic value of these conifers – and the threats they face from timber cutting and land use intensification – prompted formation in 1980 of the Central America and Mexico Coniferous Resources Cooperative (CAMCORE). The cooperative is composed of government forestry agencies, private companies and universities. CAMCORE's goal is to collect representative samples of seeds from all Mesoamerican conifers for storage in gene banks and use in breeding trials.

To select conifer species for priority collection, CAMCORE staff work with government forestry agencies in each host country to identify species and populations not well represented in existing protected areas (Table 4.2). Of highest priority are species already reduced to remnant populations throughout their range, such as Guatemala fir (*Abies guatemalensis*).

For species that are still widespread, such as Caribbean pine (*Pinus caribaea*), CAMCORE has emphasized sampling subpopulations known to be locally threatened by land clearance, timber cutting and commercial development. In cases where an entire species is a collection priority, CAMCORE tries to sample populations from different elevations, moisture gradients and soil types, thus maximizing the genetic diversity captured in a limited sample.

Table 4.2 Conifer species collected by CAMCORE since 1980 (adapted from Dvorak and Donahue, 1992)

Species	Species range		Nos sampled	
	Area	Elevation (m)	Provenances	Trees
Abies guatemalensis	S Mexico to Honduras and El Salvador	1800–4000	3	120
Pinus ayacahuite	S Mexico to Honduras and El Salvador	1800–3200	16	394
P. caribaea	Mexico to Nicaragua	10–750	18	1021
P. chiapensis	S. Mexico, Guatemala	150–2300	16	374
P. greggii	Central and NE Mexico	1200–2700	9	169
P. herrerae	W Mexico, Chihuahua to Guerrero	1200–2400	4	160
P. leiophylla	NW, Central and SW Mexico	2000–2500	11	309
P. maximinoi	Central Mexico to Nicaragua	700–2400	23	785
P. oocarpa	N Mexico to Nicaragua	250–2400	7	66
P. patula	Sierra Madre Oriental, Mexico	1500–3100	22	510
P. pringlei	SW Mexico (Guerrero, Michoacan, Oaxaca)	1500–2500	7	167
P. radiata	Central California coast	10–300	3	90
P. tecunumanii	Disjunct populations, Chiapas, Mexico, to Nicaragua	440–2770	45	1379
P. teocote	N Mexico to Guatemala	1000–3000	3	90

At a given stand, collectors gather seeds from between 20 and 50 trees, selecting individuals that display superior timber tree qualities (large size, straight trunk, healthy appearance). Where the stand is large, at least 100 m spacing is maintained between collection trees, to maximize the likelihood that seeds from different trees will have different parentage. Often, CAMCORE collecting teams are barely one step ahead of the timber cutters. An estimated 40% of the 6400 trees sampled by CAMCORE for seeds since 1980 have already been cut down by logging operations.

The conifer seeds collected by CAMCORE are distributed to member institutions for off-site storage and field trials in tree breeding programs. Although CAMCORE has functioned strictly as an *ex situ* conservation program, it has generated a wealth of information on patterns of genetic diversity within Mesoamerican conifer populations. This information will be useful for guiding *in situ* efforts for these valuable resources, such as identifying which protected areas hold genetically unique populations of *Pinus* species, or pinpointing which populations occur entirely outside of presently designated reserves [68, 69].

collectively represent a powerful set of information-gathering techniques at the disposal of resource managers charged with conserving wild plant resources *in situ*. For any particular situation, we encourage researchers and managers to consider as wide a combination of inventory methods as possible, given available resources. Each will yield unique information that may go overlooked with any single technique.

4.2 Ecological criteria for identifying priorities in plant resource conservation

Inventories can provide a wealth of information essential for managing plant resources. But to establish management priorities among useful plants, it is also necessary to have criteria for comparing and weighing their relative importance and imperilment. Take, for example, several wild relatives of oats (*Avena*) and wheat (*Aegilops*) that readily colonize roadway margins [107]. While these plants are important phytogenetic resources highly valued by plant breeders, their abundance and ability to persist in disturbed landscapes makes them lower priorities for conservation attention than other crop relatives more sensitive to human disturbance. In other situations, resource managers may have to balance attention given to useful plants with that received by other flora and fauna in a habitat or protected area. In the following pages we discuss four ecologically based criteria useful for winnowing out conservation priorities from among useful plants and their co-occurring flora and fauna:

- rarity;
- mutualists of rare and culturally important species;
- importance for maintaining the health of ecological communities;
- invasiveness.

4.2.1 Rare species

The term **rare** is usually applied in reference to a species that is scarce and difficult to encounter, perhaps due to its depletion by anthropogenic factors. Not surprisingly, we often think of rare species as being the highest conservation priority. However, the phenomenon of rarity is actually composed of three distinct ecological characteristics [184]:

- the geographic distribution of a species;
- its degree of habitat specificity;
- the size of its local populations.

As Table 4.3 illustrates, rare species can display very different patterns with respect to these three traits. Some patterns signal that a species is particularly susceptible to extinction, while others may provide a certain degree of insulation from extinction threats.

Within this classification format, species with narrow geographic distributions,

Table 4.3 Seven different categories of rarity, as determined by geographic distribution, habitat requirements and population size (adapted from Rabinowitz *et al.*, 1986, and Meffe and Carroll, 1994)

Local population size	Geographic range			
	Large		Small (endemic)	
	Habitat requirements			
	General	Specific	General	Specific
Somewhere large	Common • guava (*Psidium guava*)	Wide-ranging but only abundant in specific habitat • aguaje palm (*Mauritia flexuosa*)	Limited range but abundant in several different habitats	Limited range but abundant in specific habitat • perennial teosinte (*Zea diplo-perennis*)
Everywhere small	Wide-ranging in different habitats, but always sparse • cocobolo (*Dalbergia retusa*)	Wide-ranging but only occurs in specific habitats, and always sparse	Limited range and always sparse, but found in several different habitats	Limited range, habitat and population size (classic rare species) • Texas wild-rice (*Zizania texana*)

often known as **endemic** or **micro-areal** species [185], generally face much greater danger from extinction threats than species with wide geographic distributions. A micro-areal species can be defined as one endemic to an area of 100 km × 100 km or less. Many micro-areal species are known from one river basin, an isolated mountain, or a single island. The extinction of entire plant species becomes a very real possibility when habitat conversion or intensive harvesting impacts a species across its entire range. The late botanist Al Gentry [85] documented how at least 38 plant species apparently endemic to one small, 20 km^2 Andean foothill ridge in western Ecuador disappeared when the vegetation of the ridge was cleared for agriculture. Micro-areal species also tend to have fewer sources for recolonization if a population dies out at one site.

Whether or not a micro-areal species is ecologically limited to a specific habitat type is also an important distinction. Species restricted to a specific habitat such as gypsum dunes, serpentine outcrops or vernal pools are likely to be more vulnerable to disruption of established ecological patterns in a landscape – say, from large-scale deforestation, pollution from industrial development, or an exotic species invasion.

For plant resources that are widely distributed, the major concern facing resource managers may not be preventing wholesale species extinctions, but rather stemming the erosion of local genepools. For phytogenetic resources, such erosion can lead to the loss of genes that provide traits unique to particular subpopulations of a species.

Avocados (*Persea americana*) rank as one of the most important tropical and subtropical tree crops, worth millions of dollars annually in worldwide international markets. They are also a valuable home garden crop for rural residents throughout the tropics. The original range of wild avocados is from central Mexico to northwest Colombia, where the species appears to have been domesticated independently in at least three subregions. Wild avocados and several near relatives (*Persea schiedeana*, *P. nubigena* and *P. floccosa*) represent an enormously valuable phytogenetic resource for maintaining disease resistance, hardiness and other qualities in cultivated avocado varieties [213].

Persea americana and *P. schiedeana* in particular display wide ecotypic variability and occur in a range of habitats from frost-prone areas of the central Mexican highlands to the hot, humid Caribbean lowlands of Colombia. Neither species is currently threatened with extinction and representative populations of each species are present within a number of established protected areas, such as Mexico's Pico de Orizaba National Park and Costa Rica's Volcan Irazu National Park. Nevertheless, land across the geographic range of both species has been subject to widespread deforestation and intensive agricultural conversion. As a result, many wild avocado populations are severely threatened at the local level [213]. The cumulative regional erosion of the wild avocado genepool makes careful management of wild avocado populations in protected areas all the more important.

4.2.2 Mutualists of priority species

A single-species approach to conservation is not sufficient to ensure the survival of the many plants that depend upon a network of mutualistic relationships with other species. *In situ* conservation must aim instead to safeguard assemblages of species that are linked together by mutualisms or other critical ecological relationships.

For instance, durian (*Durio zibethinus*) is a tall, highly prized, cultivated tree that produces 'unquestionably the premier market fruit in Southeast Asia' [167, 112]. Malaysian durian stands appear to be pollinated exclusively by fruit bats of the genus *Eonycteris*, also known as flying foxes. Yet *Eonycteris* bats are not limited to durians in their foraging. Instead, as Charles Peters relates:

> These bats ... feed preferentially on the flowers of *Sonneratia alba*, a coastal mangrove which occurs in dense groves and produces a few large flowers continually throughout the year. In order to forage on this reliable food source, the bats fly 20 to 40 kilometers from their roost each night. During this journey to the coast, any durian trees that they may encounter are pollinated as a dietary afterthought.

Tragically, many mangrove stands in peninsular Malaysia are being destroyed by coastal development and it is unclear whether the bat populations will be able to survive on alternative nectar sources. Thus the viability of Malaysian durian cultivation may ultimately depend upon a mangrove and the fruit bats it supports.

Effectively safeguarding ecological mutualisms is one of the greatest conservation challenges because researchers have only just begun to study and investigate most mutualistic networks. This is particularly true for tropical and arid regions. Not only do they contain the species assemblages about which we know the least, but they also hold the ecological communities in which plants rely most heavily upon mutualists for pollination, dispersal and survival. Even intensive scientific study is unlikely to reveal all details of the complex relationships between species in a community. But by reviewing what is known about a plant resource, and thinking about where mutualisms are likely to occur (Box 4.6), managers can begin to assess species relationships important for management.

Box 4.6 Key mutualisms for management

Important mutualists for plants are often found in the following categories.

- **Pollinators**, including insects, birds, bats, certain arboreal mammals such as opossums and lemurs, and even a few lizards [42].
- **Seed dispersers**, including mammals, birds, ants and fish (particularly in the Amazon basin), that transport, drop or ingest fruits and their seeds, thereby facilitating the establishment of new populations.

- **Overstorey plants**, such as nurse trees, that are critical for providing suitable microhabitats for seed germination, seedling growth and protection from trampling or consumption by herbivores.
- **Microbial symbionts**, such as mycorrhizae and nitrogen-fixing bacteria, that enhance plant growth through nutrient uptake.
- **Organisms that provide defense** for a plant, such as ants which protect a plant's foliage and fruit from herbivores and seed predators.

Ecological studies of the life history traits of priority species (discussed further in Chapter 5) is one approach to understanding relationships between species. In addition, long-time residents in a region who are particularly observant of plants and animals may be able to offer insight into mutualisms involving useful plants. For instance, wild chile peppers (*Capsicum annuum*) in northern Mexico and the southwestern United States are dispersed predominantly by mockingbirds and thrashers (*Mimus*, *Toxostoma*) and cardinals (*Cardinalis*). This mutualism was first described not by professional ornithologists but by residents of a Lowland Pima village in Sonora, Mexico. Similarly, further south in Mexico, the Tzotzil Mayan name for the sulphur-bellied flycatcher (*Myiodynastes luteiventris*) is the same as their word for chiles and refers to its dispersal of chile seeds. These folk traditions suggest that these particular mutualisms have been observed over many generations of people, even though they may vary from region to region [148].

4.2.3 Species critical for the health of biological communities

Many ecological communities contain species that provide **connectivity** or biological links between species within food webs or microhabitat mosaics. Such **keystone** species tend to be inordinately important for maintaining species diversity within communities. For instance, the leguminous tree desert ironwood (*Olneya tesota*) is a keystone habitat modifier in the Sonoran desert. Researchers Joshua Tewksbury and Christian Petrovich [224] determined that there are 35% more species and 45% greater plant density associated with ironwood than found in random plots in the same area (Figures 4.8, 4.9). When ironwood is overharvested for firewood or the commercial trade in wood carvings, the health and reproduction of rare cacti are negatively affected because they depend upon favorable microhabitats provided by ironwood, such as shady cover from intense sun and buffering from cold winter night air [151]. The net effect of failing to protect keystone habitat modifiers is the immediate loss of rare species such as freeze-susceptible cacti, and an overall reduction in species richness.

Many species of tropical forest fig (*Ficus*) are prominent keystone plant resources as well, due to their importance within food webs [223]. Because of

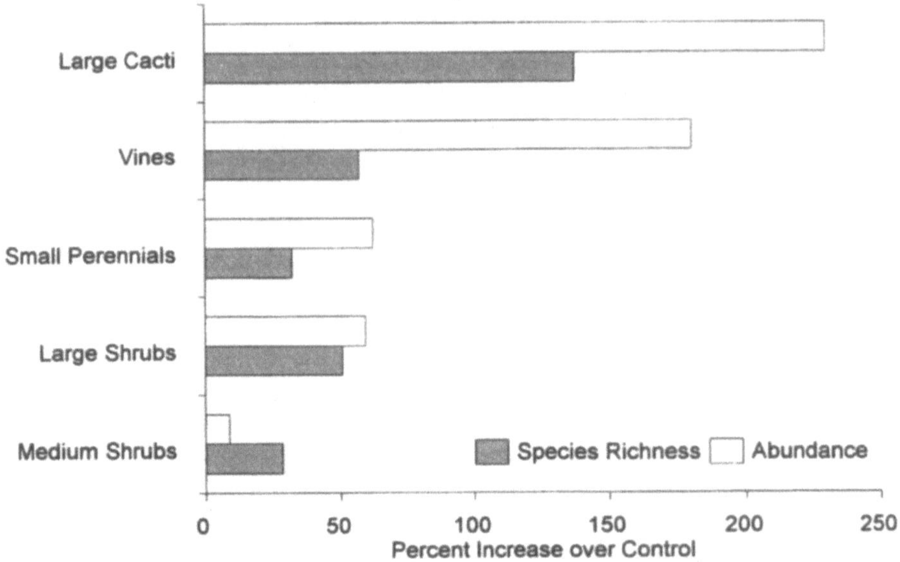

Figure 4.8 Percentage increase of species richness and abundance in perennial plant life in plots under ironwood canopies compared with random (control) canopy plots. In total, 144 canopy plots were sampled. (From Tewksbury and Petrovich, 1994, reprinted by kind permission of Conservation International.)

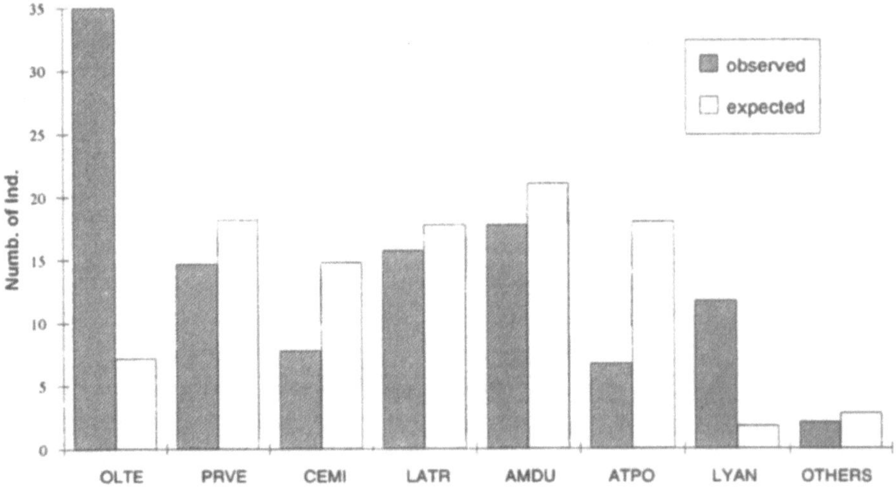

Figure 4.9 Comparison of 113 observed and expected nurse plant frequencies according to the relative abundance of each potential nurse species in the plant community. OLTE, *Olneya teosota*; PRVE, *Prosopis velutina*; CEMI, *Cercidium microphyllum*; LATR, *Larrea tridentata*; AMDU, *Ambrosia dumosa*; ATPO, *Atriplex polycarpa*; LYAN, *Lycium andersonii*. (From Nabhan and Suzan, 1994, reprinted by kind permission of Conservation International.)

their large size (many are canopy trees) and tremendous year-round fruit production, figs are a key food source for as much as 80% of the mammal species in any forest (Table 4.4). If fig trees were to disappear from a forest due to selective logging, introduced disease, or other pressures, many primates, frugivorous birds and other animals dependent upon figs would be threatened via the phenomenon referred to as **linked extinction**.

Table 4.4 Value of figs (*Ficus*) as keystone species at Manu National Park, Peru (adapted from Terborgh, 1986)

Fig species	Selected fig-dependent frugivores	Comments
Ficus erythrostica *Ficus killipii* *Ficus perforata*	Nearly all monkeys (tamarins capuchins, squirrel monkeys, spider monkeys, titi monkeys) Marsupials Procyonids Guans Trumpeters Many passerine birds (toucans, tanagers)	Nearly 60% of the frugivore biomass at Cocha Cashu relies upon figs during times of general fruit scarcity – Overall, about 12 keystone plant species, less than 1% of the plant diversity at Cocha Cashu, sustain frugivore populations during three dry-season months every year.

4.2.4 Invasive species

Managers and conservationists should take care to distinguish between native and exotic or **invasive** organisms when assessing a set of species for conservation priority. Invasive species do sometimes display some of the criteria outlined above. However, this should not obscure the fact that invasives are much more likely to be major conservation problems, not protection priorities.

Take the case of several small tropical tree species in the genus *Psidium* that are native to Central and South America. These trees produce the tasty, sweet fruit known as guava, relished by people worldwide. By any measure, guava is a culturally important plant, yet it also readily escapes from cultivation. Due to its efficient use of avian and mammalian seed dispersers, guava is an aggressive invader of natural forest, often stifling the regeneration of native tree species [213]. On the island of Mauritius in the Indian Ocean, invasion by strawberry guava (*Psidium cattleianum*) represents a grave threat to remaining areas of native forest, the last sanctuary for a host of endemic tree species. For forest managers on Mauritius, guava is a priority for eradication rather than protection [56, 222]. Another book in the People and Plants series, *Plant Invaders* by Quentin Cronk and Janice Fuller [56], takes an in-depth look at the problems caused by invasive plant species and how they can be addressed through management.

The one instance where non-native plant resources may be priorities for conservation is when they are long-established components of traditional agricultural systems (Chapter 6), have been adapted and selected as genetically distinct folk varieties, and show no propensity to escape into natural habitats and crowd out native plant communities. This is the case for a class of cacao (*Theobroma cacao*) varieties known as *criollo* cacaos. (Cacao, as many will recognize, is the Latin American tree from whose seeds chocolate is derived, and *criollo corriente* is the term widely used for folk varieties or heirloom landraces of plants.) *Criollo* cacaos (Figure 4.10) produce a chocolate of exceptional flavor and quality and had been introduced and cultivated throughout the tropics by the 18th century. However, because *criollos* are low-yielding and susceptible to fungal and bacterial diseases, they were widely replaced by other disease-resistant cacao varieties developed by breeders in the early 1900s. *Criollo* cacaos are not well represented in *ex situ* crop genebanks, and crop genetic resource specialists consider them a conservation priority worldwide wherever they occur [213].

4.3 Characterizing threats to plant resources

In comparing and prioritizing species for management, it is also important to consider the range of threats that a plant or biotic community is facing. The decline of a plant population or species typically involves both **extrinsic** or external factors (such as habitat conversion by human activities or long-term climatic change) and **intrinsic** biological factors (such as natural reproductive rates or habitat specificity), connected through often complex interactions. These interactions can be clarified by examining the threats that figure prominently in the decline of plant populations and eventual extinction of species.

Of course, extinction can be an entirely natural process, such as when a predator colonizes unaided a previously isolated oceanic island hosting vulnerable prey species. In today's world, however, extinction factors are predominantly linked directly or indirectly to human activities, and extinction rates are currently much higher than presumed natural background levels [99, 136]. To aid an understanding of how plants and other natural resources are affected by different kinds of threat, the following three divisions are helpful:

- taxon-specific threats vs. habitat-specific threats;
- proximate vs. ultimate threats;
- problems of small populations.

4.3.1 *Taxon-specific vs. habitat-specific*

Taxon-specific threats directly target a particular population or species, or a class of functionally similar species (for instance, medicinal herbs or fine hardwood timbers). Widespread taxon-specific threats affecting many plants today include harvesting pressure for commercial trade as medicines, horticultural items, food and timber; and competition and predation from introduced exotic species. For

Figure 4.10 *Criollo* cacao variety, growing in a field germplasm collection in Itabuna, Bahia, Brazil. (Photograph: R.P. Guries.)

instance, a number of Amazonian palms face heavy harvesting pressure for their fruit and heart-of-palm. Some of these palms grow very large – up to 40 m tall – and can be quite difficult to climb. As a result, many collectors employ destructive harvesting methods, often chopping down an entire palm just to harvest its fruit crop. Certain populations of these palms, including *Euterpe precatoria* (palmito), *Jessenia bataua* (ungurahua) and *Mauritia flexuosa* (aguaje), are suffering significant reductions in their size and vigor [28, 167].

Other threats do not target a single species or class of similar species, but instead tend to affect a broad suite of co-occurring flora and fauna. Such threats are diffuse and systemic, and can be referred to as habitat-specific threats. For instance, coastal mangrove forests in Honduras – and in many other tropical coastal regions – are being cut and cleared to make way for shrimp aquaculture [217]. Such activities devastate not only mangrove tree species, but also many fish, crustaceans and other marine organisms that pass at least part of their lives in mangrove communities. Climate change due to human-caused perturbation of the earth's atmosphere is perhaps the ultimate habitat-specific threat, for it has potential to disrupt ecological communities across the globe [32]. While solving the problem of climate change is a task for all of industrial society, it is not too early for natural resource managers to begin thinking about the implications of changing temperatures or precipitation patterns upon the plant resources or ecological communities with which they are concerned.

Of course, taxon- and habitat-specific extinction threats interlink through mutualisms, predator–prey interactions and other ecological relationships. A direct, taxon-specific threat to a keystone species may be felt by other co-occurring organisms as a habitat-specific threat. For instance, the charcoal industry along the US/Mexico border impacts ironwood (*Olneya tesota*) and mesquite (*Prosopis* spp.) trees directly when they are cut and burned to produce charcoal. Yet this industry also indirectly threatens 12 species of cacti and 112 other plant species which depend upon ironwood and mesquite as nurse plants [224]. These other plants are not used to make charcoal, but the loss of overstorey trees markedly lowers their growth rates and survivorship [219], thereby increasing their susceptibility to local or regional extinction.

It is even possible for a single beleaguered species to confront taxon-specific and habitat-extensive threats simultaneously. In addition to facing habitat degradation through loss of nurse plants, Sonoran desert cacti populations are also hurt directly by the illegal removal of cacti from wild habitats for the horticulture trade [22]. In clarifying the effects of such threats, managers and researchers may focus as much upon social and economic pressures as upon ecological processes (Table 4.5).

4.3.2 Proximate vs. ultimate

A second important distinction is between proximate and ultimate threats to

Table 4.5 Threats cited in studies of plant resources of the US/Mexico Borderlands (adapted from Nabhan *et al.*, 1991)

Type of threat	No. of times cited	Percentage of total threats cited
Overcollection	39	22%
Damage by invertebrates	21	12%
Damage by vertebrates	7	4%
Domestic livestock	22	12%
Urbanization	11	6%
Clearing for agriculture	8	4%
Mining and oil drilliing	4	2%
Environmental contamination	1	> 1%
Fire suppression	6	3%
Climate change	1	> 1%
Off-road vehicles	8	4%
Vandalism	1	> 1%
Changes in river flow	2	1%
Flooding (especially by dams)	6	3%
Groundwater depletion	13	7%
Loss of pollinators or dispersers	3	2%
Loss of nurse plants	3	2%
Narrow endemic	13	7%
Small population size	9	5%

plant resources [142]. In the region of Kuna Yala in eastern Panama, homeland of the Kuna Indians, one of the most valuable wild plants is a palm called *weruk* (*Manicaria saccifera*). *Weruk* leaves are the primary material used in thatching the roofs of houses, and according to Kuna botanist Heraclio Herrera, this palm has been declining in abundance throughout the region [236]. The proximate threats to *weruk* are overexploitation of individual plants for their leaves, and conversion of the palm's low-lying floodplain habitat to farm plots for growing rice and other crops. However, the ultimate threats are regional population growth, creating a demand for additional housing and thus *weruk* thatch, and the increased trend among Kuna in growing crops not only for their own subsistence, but to generate cash as well.

A management approach that takes into account both proximate and ultimate threats will have a greater chance of success than approaches directed solely at proximate issues. For instance, attempts by Kuna resource managers to place remaining *weruk* populations off-limits to all harvesting would likely be met with strong resistance from residents who quite literally rely upon the palms for the roof over their heads. Approaches that acknowledge ultimate threats such as growing local demands for housing and monetary income, and attempts to link these issues more closely to a resource's conservation, are likely to be more viable and achievable. Interestingly, some Kuna have taken this latter sort of approach on their own to address the problem of declining wild populations of *weruk* near their communities; they are promoting the cultivation of *weruk* in planted stands in agricultural areas. The palms tend naturally to grow in dense, monospecific stands and generally do well under the farm conditions prevalent in Kuna Yala. Moreover, the income that can be received by farmers selling *weruk* leaves makes it a worthwhile cash as well as subsistence crop. By looking for ways to address ultimate as well as proximate threats, managers and conservation professionals can demonstrate to local communities that they are concerned about residents' welfare and thereby serve as allies rather than adversaries.

4.3.3 Problems unique to small populations

Examining human-related threats is likely to identify the most immediate reasons why a population of useful plants is in trouble. When a population declines to a small size or occupies only a fraction of its former range, attention must also be given to potential harm from three other sources: natural **environmental** fluctuations and normal **demographic** and **genetic** fluctuations within a population. Environmental fluctuations, or **stochasticities**, are random but potentially deleterious natural events (Figure 4.11) that may affect a population and cause a

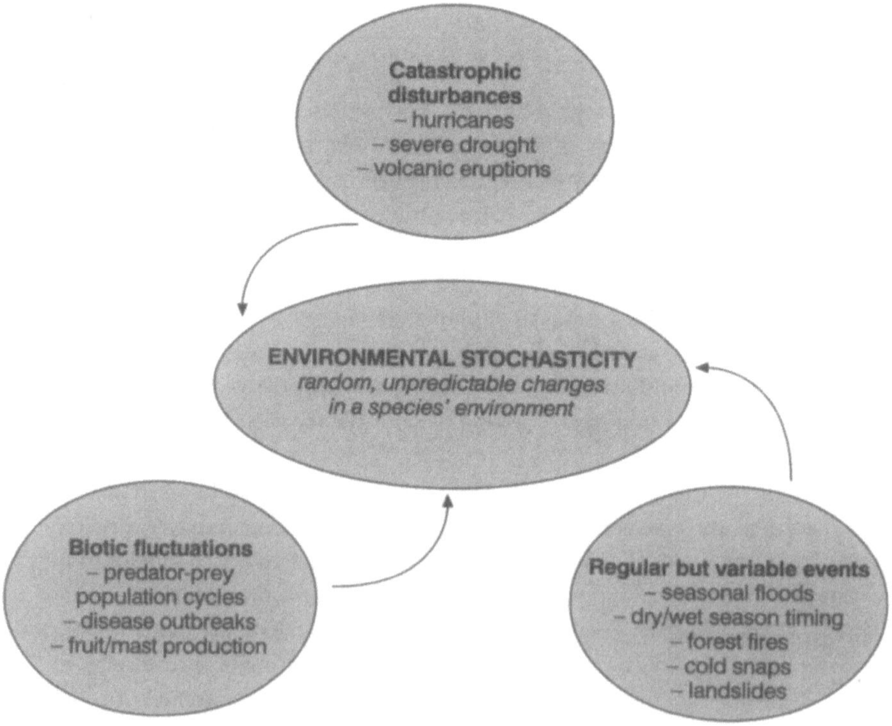

Figure 4.11 The components of environmental stochasticity.

reduction in numbers, but do not normally result in permanent harm [180]. Small populations, however, may lack the numbers to recover from such events and thus be vulnerable to extinction.

The natural random genetic and demographic fluctuations of a population can also create problems when individuals are few in number. Such problems include **genetic drift**, which leads to the loss of rare alleles; and matings between related individuals, or **inbreeding**, which can result in reduced fertility, suceptibility to disease and other genetic problems in offspring [87]. Conservation biologists have estimated that population sizes of several hundred individuals or greater are sufficient to insulate a population from most genetic and demographic fluctuations [180]. These phenomena do not automatically doom all small populations. Certain plants, particularly endemic island or mountaintop species, are known to have persisted at remarkably low population sizes for extended periods. They often have displayed low founder effects and adaptations to low breeding pool sizes, such as **autogamy** (the ability to self-pollinate) or **facultative apomixis** (the ability to reproduce asexually). However, such phenomena do put small populations at greater risk than larger ones.

How does one determine for a particular species when a population size is too small to persist on its own? Biologists have attempted to answer this question with the concept of minimum viable population size. A **minimum viable population (MVP)** is the smallest population that will likely maintain itself for a given length of time without intervention – positive or negative – by humans. A commonly used baseline size for a minimum viable population is one that will give a population a 95% chance of persisting for at least 1000 years [139]. This size will of course differ from species to species, for it is a function of each species' unique life history traits, demography and ecology.

For plant resources that are already threatened or endangered, an MVP estimate can serve as a safe target level for species recovery. For plants not yet at critically low numbers, managers can use an MVP size as a minimum level needed to prevent genetic erosion and allow continued evolution of the species. The concept can be taken one step further by examining a species' habitat requirements and the successional dynamics of that habitat (Chapter 5). With this information, MVP estimates can be translated into minimum area requirements for a population [139]. Such requirements may ultimately be more useful to natural resource managers, as it suggests a minimum amount of habitat that must be maintained to protect a species.

The big weakness of the MVP approach is that, in most cases, assembling the necessary demographic and ecological information to determine a species' MVP involves intensive research over several years, and also requires computer facilities for population modeling. Thus a complete analysis of viable population size or minimum area requirements usually presupposes extensive collaboration with universities and other research centers and will likely remain practical for only the highest priority species.

4.4 Putting it all together: planning for *in situ* management

So far in this and the previous chapter we have introduced a number of tools and concepts for surveying plant resources and comparing their relative priority for conservation. Now we are ready for the next step: to take an identified set of conservation priorities and plan for their management. A general process for developing a management plan is summarized step by step in Figure 4.12. It begins by formally ranking priorities from among a set of plant resources, and developing management **goals** and **objectives** for each one. In developing these, consideration is given to both the **external** setting in which a management effort is contemplated, and the **internal** resources of your professional management team, collaborating communities and other participants. Next, a set of **interventions** or activities is developed to achieve identified objectives. The results of all these steps are then documented and written up as a formal *in situ* management plan. You can use this planning process to identify and formalize specific management interventions wherever you are contemplating how to conserve plants or other natural resources.

Before describing further the steps in this process, let us first examine the setting in which planning occurs. Natural resource management plans are nothing new, but have commonly been prepared amid a top-down institutional setting with little external input into the planning process. This all too easily results in plans that are incomplete, unclear, overly ambitious and ultimately not very workable [197]. One alternative arrangement that promotes a more open exchange of ideas and elicits diverse opinions involves convening a workshop or series of workshops for planning. These workshops may include representatives from government agencies, local communities, non-government organizations, university researchers and other groups with a stake in a protected area or other locality.

There are a number of advantages that workshops offer for planning *in situ* management. Setting aside time specifically for a workshop gives value and importance to the planning process. By providing a more relaxed and focused atmosphere than the regular work routine, a workshop setting also encourages participants to concentrate their efforts and produce results (in this case a management plan) within a relatively short time period. Experienced organizers suggest that the ideal number of participants for a planning workshop is approximately five to 15 people. If it is necessary to include more individuals, the workshop can run more smoothly by breaking participants up into smaller discussion groups for the various workshop activities, and then reconvening to present findings to the whole group for acceptance. Many of the guidelines in the protocol for holding a community meeting (section 3.2.7) are also useful for organizing and running a planning workshop.

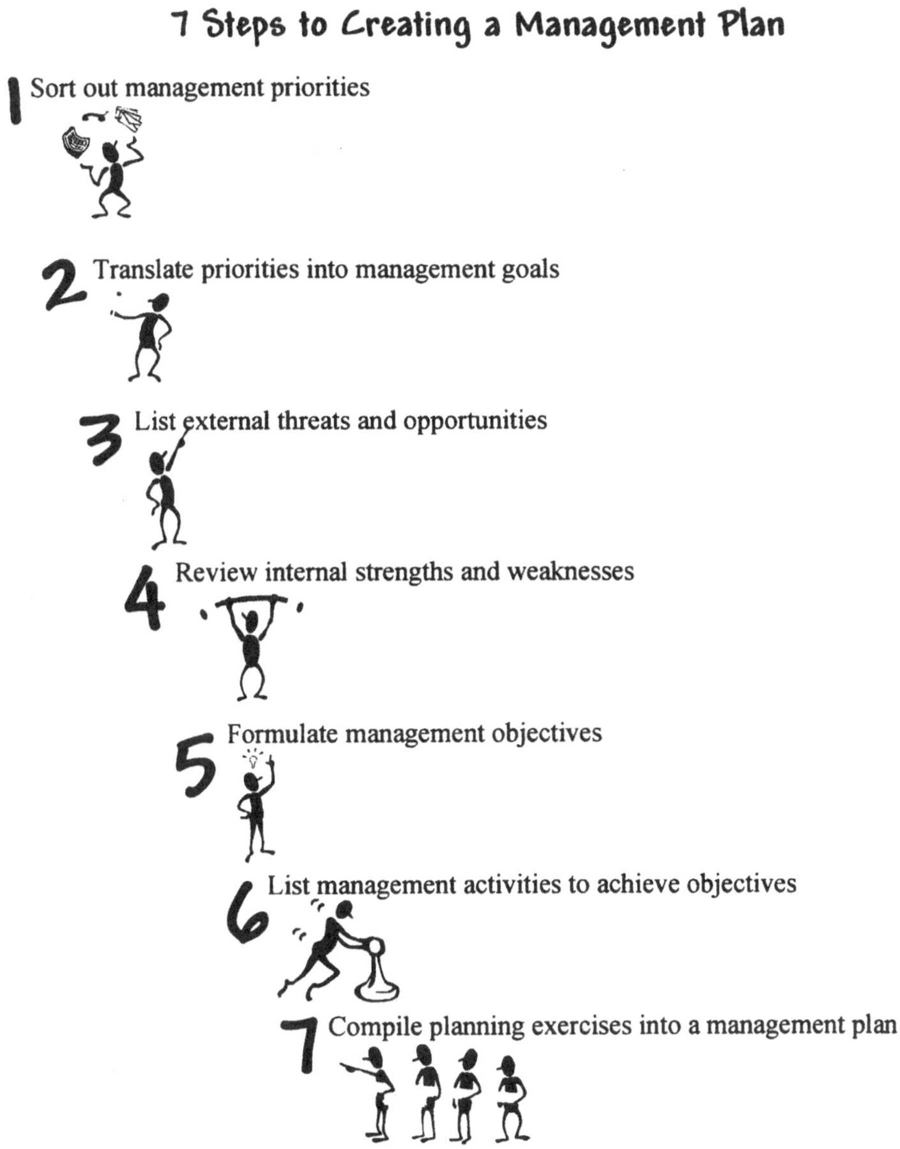

Figure 4.12 Seven steps to creating a management plan.

Workshops do not achieve success automatically, or in all situations. Prior groundwork must be laid to establish a base of trust and open up lines of communication between all parties involved. This is particularly true if there is a history of mutual distrust between protected area managers and local residents, or if serious conflicts exist between different factions within a community. The priorities which local residents and managers have for managing plant and other natural resources may be far apart at first, with little common ground evident. One way to lay the groundwork for a workshop might be to sponsor a regular series of informal community meetings attended by protected area managers to discuss natural resource issues jointly in a more relaxed setting. Such meetings could gradually build up to a joint planning session. Alternatively, a participatory rural appraisal exercise might produce a community action plan that could then be used as a basis for further planning with government resource managers and other players.

In certain situations, it may be possible to recruit experienced facilitators to guide a workshop. Conservation and development organizations, in particular, often have expertise in workshop facilitation. Outside facilitators can make the planning process less biased, and by guiding discussions can make it easier for all workshop participants to provide equal input and express opinions. In addition, facilitators can also help with the workshop's organization, keep the discussion running smoothly and ensure that the results of planning are properly documented and recorded. Facilitators do need to be selected carefully, however, because they will be in a position to influence greatly the productivity, openness and success of the entire workshop.

While *in situ* management should rest on a foundation of cultural and ecological knowledge about plant resources, it also involves taking actions on the basis of incomplete information. Planning for management is a valuable activity even without a complete inventory of the plants of an area or exhaustive documentation of all threats present. For instance, the planning process can help to determine where gaps exist in knowledge of a given protected area or suite of useful plants, and how management efforts can help to fill those gaps. Relatively few protected areas have detailed lists of the medicinal herbs, wild crop relatives and other plants which occur there. Land managers may have a general idea of the human activities affecting a region, but are less certain about how these threats affect particular species. Planning for management can still be useful regardless of how much information is available on plant resources.

As we go through the steps in this planning process, the issues involved are illustrated with examples from a hypothetical protected area network in our corner of the globe, the semi-arid uplands along the US/Mexico border between Arizona and Sonora (Box 4.7). This case study does not correspond to any one specific protected area, but is in essence a composite sketch, drawn for illustrative purposes, of major issues facing natural resource managers and others concerned with useful plants in this region. We will follow this sketch through the sequence of steps in the adaptive management process, where it is highlighted in the indicated boxes.

4.4.1 Step 1: sort through management priorities

The first step is to clarify which plant resources, key habitats, or related ecological features should be priorities for management. Where current information

Box 4.7 Borderlands biosphere reserves

Here is an initial picture of the Sonoran Borderlands Biosphere Reserve Network that might be gained from short reconnaissance visits to the area and review of secondary sources, such as regional academic studies and government survey information. This network of protected areas is distributed across 4.6 million hectares of diverse Sonoran Desert habitats, ranging from arid thornscrub–cactus forest to wetlands where the Colorado River reaches the Sea of Cortez. At slightly higher elevations, canyons drain open-canopied semi-arid woodlands. A number of canyons have spring-fed intermittent streams, which are the main permanent water sources for most reserves in the network. Some of the individual protected areas are quite large – nearly a million acres (c. 405 000 ha) in one case – while others cover only 2000–3000 ha apiece.

The biosphere reserves in the network adjoin and in some cases overlap with the traditional territory of rural villages, whose residents primarily practice ranching and small-scale agriculture on communally held lands. There are also large tracts of private rangeland held by individual landowners. In part of one reserve's buffer zone, agriculture is extensively practiced along the rich bottomlands of a large river, where several landowners have private farmland holdings. The reserve network begins approximately 50 miles from the nearest city, whose population has grown dramatically in the past two decades due to a boom in manufacturing and trade.

Most of the biosphere reserves in the network have not been systematically sampled for plants, but some have seen visits from botanical collectors throughout the past several decades. The record of collecting, as documented by plant specimens in regional herbaria, suggests that concentrations of rare plants favor better-watered locations, such as springs, narrow shaded canyons and riparian zones. Preliminary assessments of vegetation patterns on field visits and initial discussions with local residents suggest the same distribution.

about a particular plant, habitat, or protected site is too meager to set priorities accurately, this step may begin with systematic inventory approaches. Species inventories will give a picture of the plant diversity present at a site or region. Vegetation assessments and ecogeographic surveys can identify habitats rich in phytogenetic resources. Ethnobotanical inventories will illuminate the different ways in which local residents rely on plant resources.

We want to emphasize again the importance of local residents being able to add their own voices and express their opinions on what they see as management priorities. The rural appraisal methods in Chapter 3 offer one way to accomplish this, by documenting local realities and priorities which can be integrated into a planning process through workshops and exchanges. Together, inventories and consultations with local botanical experts, individuals who harvest plant products and other community members, can identify plants potentially warranting attention as priorities. Candidates for management attention can be weighed further by considering ecologically based criteria and analyses of threats to useful plant populations or species, or additional co-occurring flora and fauna. One way of pulling together all of this information in a group setting to identify and rank management priorities is for participants to work through pairwise ranking matrices (section 3.1.6) to compare plant resources directly [132]. Another approach is the 'snow-card' technique (Box 4.8), which is designed for use in workshops or meetings, and can function well with a diverse group of participants.

Box 4.8 The snow-card technique

The snow-card technique is a tool that a group or team can use to develop and organize ideas on any topic. The method allows each group member to present their ideas and opinions, reduces bias, minimizes repetition, and ideally fosters a sense of teamwork while reaching consensus. In this example we have tailored the snow-card method specifically to identify priorities for managing plant resources.

1. Materials needed:
   – index-sized cards or slips of paper (about 6 cm × 10 cm each);
   – pens or markers;
   – tape or thumbtacks;
   – a blackboard or clear wall for tacking up cards.
2. To initiate the procedure, it may be useful if one knowledgeable person summarizes the general topic of discussion – for instance, useful plants native to the surrounding region. This allows all participants to have a more equal understanding of the subject, particularly if they come from different backgrounds. Be sure to advise the speaker to review the plants or other topics in an unbiased fashion.
3. The speaker should then ask the group if anyone else has information that may have been missed. That information should be presented as well.
4. After the opening discussion is finished, a facilitator passes out several cards to each person, with some markers. The facilitator asks each person to write down the plant species, sites or habitats (one per card) that they think are priorities for management at the site. Give each person 10 minutes or so for this step.

5. Once everyone has finished, the facilitator takes all the cards and mixes them up, then tapes each of them to the wall, reading them off while taping them up. The facilitator asks if there are any more ideas, and if so these are placed on the wall also.
6. When all the ideas are up, the facilitator works with the participants to cluster the cards in several categories. This can be done by asking participants which plant resource cards seem to go together. As each cluster is created, its general title is written down and placed as a heading.
7. The categories are not designed to rank the plants or other card subjects, but instead serve to organize the general discussion. To discuss plant resources, the following categories may prove useful for clustering:
   - home subsistence products
   - income source for local residents
   - ceremonial or medicinal plants
   - phytogenetic resources
   - rare/difficult to find
   - common/easy to find
   - mutualist
   - keystone
   - invasive
   - other.
8. Once all the item cards are clustered, the facilitator asks participants to prioritize the items. Starting with the first category, the facilitator works through the cards, asking participants which items are most important to achieve for the protected area or other site, and then fosters discussion and works with the group to get consensus.
9. As each priority is determined, the facilitator puts the corresponding number on the card – first priority gets a 1, the second a 2, etc. When all have been prioritized, they are placed in order with a heading of 'Conservation priorities for plant resources in [name of region, municipality, or protected area]'. That's it – you have your priorities list.

The end result of this first step should be a preliminary list or subset of the local flora and fauna that can be considered to be priorities for management activities within the jurisdiction of a protected area, community land, or other site. We call this list preliminary because it will be subject to evaluation, reassessment and adaptation over time. As more is learned about the biota of the region or as local use patterns change over time, new priorities may emerge, or it may become apparent that certain species do not require priority attention.

Once a priority list is developed (Box 4.9), it is time to proceed to Step 2, translating conservation priorities into management goals. At this point, you may want to follow Step 2 through to Step 5 (Figure 4.12) separately for each priority, rather than running all priorities through the steps at once. This will make the process clearer and tends to give a more accurate impression of the work involved in managing for each priority.

Box 4.9 In situ conservation priorities in the Borderlands

Through workshops and consultations with local residents throughout the region, along with reviewing regional ethnobotanical studies, we can identify several ethnobotanical products that are particularly important for local livelihoods. On foothill slopes dominated by desert scrub, there was once intensive harvesting of wild *Agave* plants for the bootleg mescal industry. Several common species of arborescent cacti, particularly those in the genus *Opuntia*, provide edible fruits, buds and stems or pads. Most wild cactus products appear to be gathered for home consumption, but they also appear in local markets. Initial discussions with harvesters suggest that the harvesting pressures on cacti are relatively low. At other lowland and foothill sites in and near several biosphere reserves, mesquite (*Prosopis*) and ironwood (*Olneya*) trees are commonly cut for fuelwood and charcoal production. Villagers indicate that while they primarily harvest trees only for home use, commercial charcoal producers have begun cutting in their area. In several villages, residents carve ironwood into figurines that are sold to visiting tourists, and this trade has become a major source of income for the community. In the oak-dominated *encinal* forests at slightly higher elevations, acorns are collected for home consumption and sale. In scattered locations, stands of wild chile peppers (*Capsicum annuum*) – the ancestor of most domestic chiles – are harvested seasonally for their fruits, which are sold in the spice trade.

Of these plants, agaves and chiles are also phytogenetic resources. Agaves and chiles – both wild progeny and cultivated folk varieties – are commonly planted by rural residents in backyard gardens, hedgerows and other areas around communities. Other important wild crop relatives in the region include wild beans (*Phaseolus*), cotton (*Gossypium*), canyon grapes (*Vitis*), Palmer's saltgrass (*Distichlis*) and gourds (*Cucurbita*).

What sort of *in situ* management priorities might emerge for plant resources in the Borderlands Biosphere Reserve Network once these species are compared via conservation criteria and threats? Preliminary information suggests that at least three phytogenetic resources – chiles, Palmer's saltgrass and certain agave species – show some traits of rare species, though not so restricted in range that they are micro-areal endemics. Legume trees at lower elevations display traits of keystone species, particularly in their role as nurse trees for many other plant species. Oaks may also be a keystone food resource for local wildlife, including mule deer and javelina (collared peccary). Drawing upon the research of ecologists and folk knowledge of local residents, we know that agaves rely upon a specific mutualist (long-nosed bats) for pollination. Wild chiles also depend upon birds as mutualists to disperse their fruit to preferred microenvironments for seed germination and growth. In sum, a group snow-carding exercise might identify the following management priorities among useful plants:

1. Wild chiles (ethnobotanical, phytogenetic, rare, mutualistic).
2. Agaves (ethnobotanical, phytogenetic, mutualistic, common to rare – depending on species).
3. Legume trees (ethnobotanical, keystone, common to uncommon – depending on species).
4. Oaks (ethnobotanical, keystone, common).
5. Palmer's saltgrass (phytogenetic, ethnobotanical, rare).
6. Arborescent cacti (ethnobotanical, common).

4.4.2 Step 2: translate conservation priorities into management goals

In this step, participants lay out management goals for each conservation priority (Box 4.10). A **goal** is a desired situation or accomplishment that a team works to achieve over time. It is what you plan to manage for over the long term for a priority plant species or important habitat type. Goals are usually stated broadly and kept relatively general; as the planning process progresses,

Box 4.10 Borderlands management goals

To illustrate goals, let's take one of the priority plants in our Borderlands case study, wild chile peppers. For chiles, our initial management goal might be:

> To conserve wild chile populations within the Sonoran Desert Borderlands Biosphere Reserve Network for protection as a genetic resource and for direct utilization as a locally harvested product in a UMA, a Mexican area of management and extraction.

How can we tell if this goal is realistic? The best test is to develop objectives and activities for this goal. If we have a hard time identifying workable activities, then we may need to revisit our goal and consider changes to make it more workable.

they will become supported by specific objectives and activities. A good format for setting out management goals is to present the list of identified priorities to a group of participants and have them nominate and discuss goals for each priority. It may be more efficient to assign individuals or small subsets of participants to develop a goal or goals for a single priority and then have all goals reviewed by the entire group.

4.4.3 Step 3: list threats and opportunities external to your management team that are related to the stated management goal

Without the benefit of a formal planning effort, managers and conservationists may fail to step back and take a broad look at the potential threats and opportunities that may hinder or help them in achieving their conservation goals. By taking stock of their external environment, managers can better anticipate potential problems – for instance, when commercial demand increases for a wild medicinal plant, triggering increased exploitation. Likewise, managers may take greater advantage of favorable events, such as individual farmers who, on their own initiative, continue to grow traditional varieties of crops that other farmers have replaced with modern cultivars.

Both threats and opportunities can be reviewed in the planning process by participants in workshops and group meetings (Box 4.11). After listing a goal for managing plant resources, participants can brainstorm threats and opportunities that are likely to hinder or help achievement of that goal. For instance, one goal might be to conserve a plant with strong commercial value that is being overharvested primarily by outsiders rather than by local residents. In this case, one opportunity for management might be to mobilize local communities for better protection of remaining plant populations from outside exploitation. Working closely with residents to manage plant resources may also be an opportunity for protected area managers to compensate for shortfalls in official funding or staffing.

Box 4.11 Borderlands threats and opportunities

Here are presently identified threats and opportunities that relate to our management goal: to conserve wild chile populations within the Sonoran Desert Borderlands Biosphere Reserve Network for protection as a genetic resource and utilization as a locally harvested resource.

Threats

Commercial harvesting pressure that is intensifying

Traditionally, people only harvested a small portion of the yearly fruit production. But commercial demand for chiltepines is rising with the growth of urban areas in the region. It is now increasingly common for collectors to harvest the entire annual crop of a population, sometimes by breaking off branches or uprooting chile plants [154].

Habitat loss due to agricultural conversion and intensification, and urbanization [92]

These processes predominate in lowland areas and near cities, where human disturbance is more intensive. Proximate habitat-related threats are less of a problem for chile stands in upland regions with fewer people, as in more remote biosphere reserves.

Bird dispersal of chile fruits [148]

Fluctuations in population sizes of chile-dispersing bird species – say, due to pesticide poisonings or decreased nesting success – may influence the success with which chile plants colonize or recolonize potential habitat sites following local extinction events.

Environmental limits

Chile populations in the Borderlands exist at the limits of the species' range, and experience climatic extremes of cold and aridity. As a result, individual chile plants are almost always found growing in favorable microclimates provided by nurse trees – hackberries (*Celtis*) or wolfberries (*Lycium*) – or boulder piles. During an extended drought or cold snap, chile plants growing in marginal sites may die off. In addition, certain nurse plants of chiles are not well adapted to fire, so the microhabitat they provide may be threatened by periodic burns.

Opportunities

Local concern over chile harvesting practices

Interviews with chile harvesters indicate that there is already a degree of self-patrolling under way to prevent the most egregious harvesting methods, such as uprooting entire plants. Managers might be able to support locally initiated efforts to monitor the harvest or document the ownership of chile stands to establish traditional use rights.

Value of wild chiles as a genetic resource

Their value might enable managers to obtain funds from international agricultural agencies and other sources to support research on the ecology, distribution and other aspects of wild chiles. The new Mexican government management unit, UMA, fits this value.

Cultural interest in chiles

The prominence of chile peppers in regional cuisine, folklore, celebrations and other aspects of Borderlands cultures offers fertile ground for generating broad public support for conserving populations of wild chiles.

4.4.4 *Step 4: identify internal strengths and weaknesses for* in situ *management*

Recognizing the current strengths and weaknesses of one's organization, agency or other institution can help to ensure that the goals and objectives set out for managing plant resources remain realistic and attainable (Box 4.12). This step encourages the management team to take stock of its internal resources:

- **human** (staff and volunteers);
- **physical** (equipment and supplies);
- **financial** (funds);
- **reputation** (especially in terms of credibility with local residents).

Analyzing strengths and weaknesses will aim the project or team in the right direction and highlight the areas where improvement is needed. While assessing strengths and weaknesses tends to be a more self-reflective exercise than other steps in the planning process, it can often be accomplished through similar group reviews and brainstorming sessions.

4.4.5 *Step 5: formulate management objectives that minimize threats and weaknesses and maximize opportunities and strengths*

Objectives are specific short-term accomplishments designed to achieve part of a

Box 4.12 Borderlands management strengths and weaknesses

Strengths of the management team working at sites in the Biosphere Reserve Network include:

- a close working relationship with a cadre of dedicated biologists at regional universities and museums, who are intimately familiar with the ecology, habitats and species of the Borderlands;
- a dedicated and energetic field staff with a strong awareness of the need to work with local residents;
- a good reputation of the management team among local residents and landowners, based on past experiences;
- new designation, UMA, fits species needs.

Examples of weaknesses include:

- government resource managers overworked;
- limited administrative budget, with few prospects for expanded government funding in coming years;
- goals of management team conflict with those of other government agencies in the region, such as efforts to boost agricultural productivity by providing access to cheap pesticides that harm pollinators of wild plants;
- no strong links yet with anthropologists or social scientists who could provide advice on social aspects of management activities.

more general long-term goal. The process of formulating objectives provides a check to ensure that identified goals are realistic, and offers a better sense of how much work will be involved. As with goals, a good format for laying out management objectives is a group meeting or workshop where a facilitator assigns individuals or small groups to develop objectives from goals. The objectives are then reviewed and debated by the entire group for their relevance and workability.

It is also worth pointing out several subtle aspects of objectives that influence their usefulness. Objectives that are poorly worded, vague, inoperable or conflicting will be of limited guidance. Some criteria that can help to identify whether an objective is appropriate and well stated are as follows.

- **Specific** – describes a single key result to be accomplished.
- **Measurable** – readily monitored, can tell if it is accomplished, quantifiable if possible.
- **Result-centered**, not activity-centered.
- **Concise** – to the point and understandable.
- **Realistic** and **substantial** – neither too hard nor too easy to accomplish.
- **Consistent** with other objectives.
- **Time-bound** – where possible, specifies a date for accomplishment.

Having analyzed the external threats and opportunities and the internal strengths and weaknesses leaves the management team well prepared to develop achievable objectives (Box 4.13). Be sure to use that information when forming the objectives.

4.4.6 Step 6: list possible management activities to achieve objectives

It is at this stage that participants in the planning process identify, debate and sketch out specific management activities. Once these activities are written down and agreed upon, they represent a road map for meeting *in situ* management objectives for useful plant resources. To be effective and relevant on the ground, activities should address objectives directly, keeping within the threats, opportunities, strengths and weaknesses outlined in the previous step. Sample management activities for wild chiles in the Borderlands are presented in Box 4.14. In addition to listing activities, the team should also determine who will be responsible for carrying out that activity and develop a timeline for accomplishment.

Once these five steps (goal/threats and opportunities/strengths and weaknesses/objectives/activities) have been run through for the first priority, the process can then be repeated for the next one. Of course, many threats, opportunities, strengths, etc. may be the same for more than one priority. Management activities should include ways to monitor the progress of *in situ* management as it is implemented, as well as ways to evaluate whether the

Box 4.13 Borderlands management objectives

Let us return to wild chiles to illustrate objectives for our identified management goal.

Goal

To conserve wild chile populations within the Sonoran Desert Borderlands Biosphere Reserve Network for protection as a genetic resource and utilization as a locally harvested resource in a Mexican government-designated Unidad de Manejo y Approvechamiento (UMA).

Objectives

1. To maintain the current extent and numbers of wild chiles within designated protected areas.
2. To sustain current chile nurse plant densities within designated protected areas.
3. To increase the use of ecologically sound harvesting approaches among at least 50% of local residents harvesting chiles within or near protected areas within two years.
4. To develop a travelling educational exhibit within 2½ years to improve public awareness of chiles as a resource.

Box 4.14 Borderlands management activities

Objective	*Activity*
1. Maintain the current extent and numbers of wild chiles within designated protected areas and in newly designated UMA.	Map locations of wild chile populations. Monitor persistence of all chile stands and population numbers of selected stands. Assemble, from literature and field observations, information on rates of chile fruit production and other life history traits.
2. Sustain current chile nurse plant densities within designated protected areas.	Establish hackberries and wolfberries on sites needing revegetation. Eliminate competing non-native shrubs at sites where chiles are concentrated.

3. Increase the use of ecologically sound harvesting approaches among at least 50% of local residents harvesting chiles within the reserve within two years.	Conduct rural appraisal with villagers to determine past and present extent of harvest, methods used, and constraints on adopting new approaches. Sponsor workshop to explore options for collaboration with local harvesters to monitor harvests.
4. Develop a traveling educational exhibit within 2½ years to improve regional public awareness of chiles as a resource.	Apply for funding to cover costs of developing and promoting exhibit. Sketch out ideas for contents of exhibit and meet with local educators to brainstorm on concepts. Record oral histories with residents for inclusion in exhibit. Assemble components, field test, refine. Monitor media coverage, local feedback, audience response to evaluate success of exhibit.

stated goals and objectives were met. Monitoring and evaluation activities are discussed in the next chapter – they are an important part of any management process, yet are often overlooked in planning.

In this step, planning participants will always want to assess possible management activities against available resources: finances, personnel, equipment, expertise, potential support from other institutions and so on. For every potential activity, managers should ask the following questions. Of the management objectives identified, which ones can realistically be addressed? Who will be responsible for doing each activity? Are there novel ways of dealing with identified threats that have not yet been tried by the management team? How might new resources be mobilized?

Once the initial planning effort produces a list of activities, those responsible for implementing activities can plan their work in more detail. Subsequent meetings can generate a detailed list of tasks and a timeline that will need to be done for each activity. For example, if an activity is to hold a workshop, then one task will be to prepare workshop materials. Another related task will be to arrange the workshop site and meals. Be sure that each task has a deadline and is incorporated into a general timeline.

The activities and tasks that sift out through this filter represent collectively the end product of the planning process. When done properly, they should form a well grounded, specific and realistic map of what managers, their collaborators and their co-workers will do to accomplish goals and objectives in support of management priorities.

4.4.7 Step 7: compile planning results into a management plan

In any planning process, it is important to keep a permanent record of priorities, goals, threats, weaknesses, objectives, activities, tasks and other components that make up a management plan. One approach is to designate one or several individuals as recorders or scribes at the start of each planning session. Flipcharts and large sheets of paper tacked on a wall can be used to record information during the workshop or meeting. At the close of the planning process, this documentation can then be assembled into a formal written plan. The plan can serve as a reference point when implementing management activities and as a benchmark for measuring progress. A sample plan is presented in Figure 4.13. Note that the plan should clearly identify who is responsible for implementing each activity and include a timetable for completing activities.

In addition to a planning document, the team may want to maintain information

Management Plan Contents

Part 1. Introduction to the initiative
- *purpose of project*
- *background review of project*

Part 2. Project priorities

Part 3. Project goals and objectives related to priorities

Part 4. External threats and opportunities for management

Part 5. Strengths and weaknesses of the management team

Part 6. Management activities to be undertaken
- *targeted towards project objectives*
- *ranked in order of priority*
- *includes monitoring and evaluation activities*

Part 7. Management team responsibilities
- *list of who will do which activities*

Part 8. Timeline for accomplishing activities

Part 9. Conclusions

Appendices:
- Information sources cited
- List of management team members and their qualifications
- List of agencies and organizations involved in management effort
- Maps and other illustrations
- Letters of support for project

Figure 4.13 Contents of a sample management plan.

on small posters or flip charts. For example, project goals and objectives can be posted in an office, community hall, or other central location. Such information can serve as inspiration and a daily check that things are on the right track. Also, the team may want to keep a large poster with a matrix of the goal, objectives, activities and tasks with a timeline in an area that everyone can see and update periodically. This is an informal mechanism for keeping the team up-to-speed and demonstrating accomplishments.

4.5 Summary: keeping management adaptive

Although completing a management plan is certainly an accomplishment, it should not be viewed as an end in itself or a final product. The real value of a plan is in how it is implemented to facilitate management of plants and other natural resources in a protected area. It is impossible to foresee all situations and scenarios, so any management plan is bound to have shortcomings. Some activities will be successful, while others may need revising, overhauling, or a change in emphasis. Certain objectives will be right on target while others may prove difficult to achieve on the plan's time frame. Activities to monitor a plan's impacts cannot be stressed enough, for they are the primary means of discerning how a plan is doing, what is working and what is not. Keeping management adaptive also requires that there be a way to evaluate a plan periodically and adopt changes. These monitoring and evaluation components are essential for *in situ* management, yet are often overlooked in a plan's preparation [197]. How to incorporate them into management effectively is the subject of the next chapter.

5

Monitoring and evaluating plant resource management

Figure 5.1 Ecologists Alfonso Valiente and Joseph MacAuliffe teaching use of the log plot method to range managers from Chihuahua, Mexico, during a workshop on vegetation evaluation methods. (Photograph: P. Mirocha.)

No matter how clearly a management plan for natural resources or a protected area is laid out and implemented, there will always be the question: is it working? Individuals both inside and outside a management effort need ways to judge a plan's progress as it is implemented and for adapting management practices to changing situations over time. Resource managers will often have to ask and then answer questions such as: How can we tell whether a protected area is making a difference in the status of local plant populations? Are the activities we have implemented meeting our identified goals and objectives for useful plants? What are the ultimate effects of our work on native habitats, the flora and fauna they contain and the well-being of local residents?

Monitoring and evaluation can help managers to answer such questions. These two closely related subjects are receiving increasing attention from practitioners in both conservation and community development arenas. They are explored here in the context of *in situ* plant conservation. First, we review an approach to planning for these subjects in a resource management program: how managers can identify what, where and when to monitor and evaluate various subjects. We then present in greater depth a number of tools useful for monitoring and evaluating the status of plant resources.

5.1 Defining monitoring and evaluation

Although the terms 'monitoring' and 'evaluation' are sometimes used concurrently, there is a distinction between the two. **Monitoring** refers to the regular collection and analysis of information to determine whether or not activities are working and explain why. The information gathered in monitoring can be **ecological**, such as trends in vegetation cover or numbers of a species; **physical**, for instance soil loss or nutrient recovery; **social**, such as monthly family income or time spent on particular activities; or a feature that reflects a combination of these categories, such as average fallow period in a farming region. Monitoring is usually done at regular intervals – daily, weekly, monthly, or seasonally – so that over a cumulative time period (a month, a year, or longer) trends in a particular situation become evident and measurable.

Evaluation is an opportunity for participants in a conservation effort to step back, review information and look at whether their work has met the desired goals and objectives [62]. It is a chance to take stock of the effects – positive and negative, expected and unexpected – created by a project's actions and activities. Evaluation and monitoring are often part of the same process. For instance, the information obtained by monitoring the reproductive success of a plant population can be indispensable for evaluating management designed to maintain habitat favored by the plant. Overall, evaluation tends to take place less frequently (for instance, at a project's midpoint and end) than monitoring, which is more of a continual process. Evaluation also offers an ideal opportunity for debating and modifying the direction of a project or management effort, or for addressing problems that have arisen during implementation.

Monitoring and evaluation activities can benefit management of useful plants in several ways. The information obtained through regular monitoring makes it more likely that an *in situ* management plan will attain its goals. Periodic evaluation of a project or plan also provides incentive for the management process to change over time, in an adaptive, self-correcting fashion. Conservation professionals and land managers can use evaluations as formal opportunities to identify which initiatives are working well, and why, or to find alternatives to those which are proving ineffective.

5.2 Planning a monitoring program

The best way to prepare for monitoring is to make it an integral part of an *in situ* management plan by setting out monitoring activities that correspond to each objective. Involving monitoring as part of the overall planning process gives you an opportunity to coordinate the time, staff and money needed for monitoring with other management demands.

5.2.1 Deciding what to monitor

Monitoring will be most efficient when it targets specific indicators – social, economic, ecological and physical – selected by reviewing the objectives set forth in the management plan. Box 5.1 illustrates several indicators related to management objectives for wild chile peppers in the proposed UMA, our example from the last chapter.

Appropriate and effective indicators should be **specific** rather than general, and readily **quantifiable** or measurable. They may be either **direct** or **indirect** measures of management effectiveness [62]. The indicators in Box 5.1 are

Box 5.1 Indicators for monitoring Borderlands management objectives

Objective	*Indicators*
1. Maintain the current extent and numbers of wild chiles within designated protected areas.	– Population size – Numbers of populations
2. Sustain current chile nurse plant densities within designated protected areas.	– Population size of nurse plants (plants/ha) at chile stands – Amount of habitat occupied by suitable nurse plants – Amount of habitat occupied by non-native competitor shrubs

3. Increase the use of ecologically sound harvesting approaches among at least 50% of local residents harvesting chiles within the reserve within two years.	– Percentage of harvesters who follow techniques – Percentage of harvesters who participate in oversight and self-patrolling – Amount of damage to plants evident after harvesting
4. Develop a travelling educational exhibit within 2½ years to improve regional public awareness of chiles as a resource.	– Numbers of people attending educational chile events – Numbers of presentations and educational seminars given on chiles

primarily direct measurements of the objectives. Box 5.2 illustrates how indirect indicators can be used for monitoring, in this case of **genetic erosion** – the loss of genetic diversity among wild relatives of crops and other phytogenetic resources.

Box 5.2 Monitoring genetic erosion

Researchers at the International Plant Genetic Resources Institute (IPGRI) have developed a system to monitor how gene pools of traditionally cultivated plants and wild crop relatives become reduced in abundance and distribution (94). The system can be used to compare different species in a single region, different populations of the same species, or one population at different points in time. Since it is difficult to measure genetic erosion directly, the IPGRI system relies heavily on indirect indicators of a species' susceptibility to genetic erosion. For wild species, these include the following.

- **Ecological indicators**:
  - distribution of the species or variety (rare vs. abundant);
  - extent of habitat favored by the species (restricted vs. extensive).
- **Climatic indicators**:
  - frequency of drought (two years running, more than once every 10 years, less than once every 10 years);
  - susceptibility to climate change (species restricted to summit or low-lying coastal areas).

- **Anthropogenic disturbance indicators:**
  - extent of use of the plant (industrial exploitation, local subsistence use, not used, protected);
  - agricultural pressure on habitat (large-scale cultivation, subsistence cultivation, uncultivated).
- **Social indicators:**
  - annual human population growth rate (> 3%, 1–3%, < 1%);
  - distance of a site from major population center (< 20 km, 20–50 km, > 50 km);
  - distance from major development project (< 20 km, 20–50 km, > 50 km).

The full array of IPGRI indicators for both wild and cultivated species is listed in Appendix B. The system is implemented by scoring each indicator on a 0 to 15 point scale, and then summing all scores for a species or population to give a measure of the magnitude of genetic erosion. Field tests of this system in 1987 found that it accurately predicted the level of genetic erosion (as measured more directly by loss of populations and reduction in population size) experienced by rare or local species that had relatively specific habitat requirements, such as wild coffee (*Coffea*) in the montane forests of Kenya [107]. However, the system proved less accurate for species that can withstand a degree of human disturbance or do well in marginal habitats, such as certain herbaceous wild relatives of forage legumes. Predicting genetic erosion for disturbance-tolerant species clearly requires a different set of indicators.

In addition, this system is primarily designed for tracking genetic erosion on a national or regional level, and is likely to be implemented primarily by agricultural researchers and other conservation professionals. Monitoring the erosion of crop gene pools in a single farming community would require different, more fine-scale indicators, such as the percentage of families who report a shortage of available labor to help with on-farm work (see Box 6.4). This kind of monitoring effort might be better implemented through a more collaborative approach, such as one employing regular community meetings and interviews.

5.2.2 Deciding how to monitor

For any monitoring it is best to identify at the outset which tools and skills will be needed to measure, assess and record a particular indicator. These may include staff training exercises, equipment, supplies, or other preparation. Later in this chapter we review a number of tools and assessment methods that can be put to use for monitoring plant resources. Sometimes a single technique can be used to examine several indicators. For instance, collaborative mapping might identify both the dimensions of remaining forestlands and the extent of lands severely degraded by soil erosion. In other situations, several tools may be required to verify the status of an indicator. Assessing changes in tenure arrangements over land or resources may well require interviewing, mapping and possibly field transects. At this point, it is a good idea to think again about how the monitoring activities contemplated can be accomplished with the funds and other resources you will have available. If what and how you plan to monitor do not look feasible, then you will need to consider alternative indicators or methods.

5.2.3 *Deciding who will monitor*

Identifying who will monitor plant resources depends primarily on the range of people involved in the process of planning for *in situ* management and the kinds of technical skills required for the monitoring. If the planning process has been collaborative, then monitoring can often be carried out in an equally open fashion. Some organizations and groups involved in community development work have experimented with participatory approaches where local residents, in consultation with conservation professionals, identify which variables will be monitored, the techniques to be used and other details of the monitoring process.

Even if local residents have not been directly involved in planning for management, we still encourage you to consult with community members and involve them in monitoring efforts to the fullest extent possible. One arrangement that has proven productive in some cases is for researchers or other conservation professionals to train local counterparts in the technical methods needed to monitor a plant resource trend, and then let them proceed with implementing the monitoring [235]. Regardless of whether monitoring is to be undertaken by local residents, agency staff, consulting scientists, or a combination of all three, the monitoring workplan should clearly specify who will be responsible for the various activities.

5.2.4 *Deciding where to monitor*

Once monitoring indicators have been selected, there is the question of locating appropriate study sites. In some cases, you may want to select sites **randomly** in order to minimize potential bias in the results and allow for intensive statistical analysis. In other situations it may be important to have a **systematic** sample across an entire area. These and other distinctions apply to collecting both social and ecological information and are discussed further in sections 3.1.1 and 5.4.1.

In addition, it is common for a monitoring effort to target specific monitoring sites that exhibit a desired set of qualities, such as vegetation of a certain age or habitat, or a particular land use history [132]. Local farmers, plant harvesters and other knowledgeable residents can often provide invaluable help in identifying appropriate locations for monitoring particular plant resources (Box 5.3). They can identify and illuminate distinctive features of a site's history, particularly those related to past human uses such as swidden cultivation or plant harvesting. Involving local resource users in both the design and data collection stages of monitoring allows for the input of indigenous knowledge about a plant's life history and abundance and can increase local awareness of resource conservation efforts.

Box 5.3 Identifying study sites with local residents

In 1996, conservationists from the Biodiversity Conservation Network and the Solomon Islands Development Trust wanted to examine the potential ecological impact of a community development enterprise to extract an oil from the nuts of the *ngali* tree (*Canarium indicum*). They decided to use plot surveys to compare the demographic status of ngali stands that were harvested regularly for their nuts, with those that were unharvested [239].

Consultations with residents from the communities involved in the enterprise revealed that locating an unharvested control plot would be difficult. Demand for ngali nuts was strong, and essentially all ngali stands within a three-day walk of the village saw harvesting pressure. However, residents also pointed out the existence of sacred 'tambu' sites, where many land use practices were customarily forbidden. Not all tambu sites were appropriate control sites, since some allowed collecting of ngali nuts and others contained too few ngali trees to be a representative sample. Eventually, however, local consultants identified suitable tambu groves. After discussions with local leaders and landowners about the sampling activities that would take place in the tambus, the survey team received permission to enter and lay out demographic sampling plots (section 5.4.5). At the request of local leaders, the team performed customary prayers before entering the grove, carefully obeyed all traditional rules and rituals once in the tambu area, and also paid proper respects to the site after completing the survey [239].

5.2.5 Deciding when to monitor

It is best to let the particular features of an indicator determine the optimum schedule for when and how often the indicator should be monitored. For instance, a project in Ecuador assessed three different management regimes for an economically important palm species by measuring the average growth rate of palms under each type of management. Growth in palms is measured by the appearance of new leaves over time. Accordingly, individual palms were censused every three months to make sure no new leaves were overlooked and to obtain information across both wet and dry seasons [235]. A project that used growth as an indicator to monitor management of a timber tree might only need to measure once per year, since the indicator would likely be each tree's trunk diameter, a variable that in most timber species does not change dramatically from month to month.

5.2.6 *The importance of baseline information*

In many cases, monitoring will proceed best when there is a base of knowledge about a plant resource accumulated prior to management intervention. Without such baseline information, it becomes harder to identify the effects of changes in management strategies and to assess the role of unanticipated factors, such as a particularly severe drought. There are several approaches that managers can take to obtain baseline information for monitoring. First, it is important to review secondary information sources, such as previous studies or field survey reports, for baseline data. Secondly, when management teams are ready to gather information themselves, the rural appraisal methods in Chapter 3, the inventory methods in Chapter 4 and the field methods presented later this chapter offer a good starting point. Collecting baseline information is also an area for collaboration with regional scientists skilled in survey methods and familiar with the plants, communities, or localities that will be the subject of monitoring.

5.2.7 *Analyzing and using monitoring information*

Since monitoring is designed to provide insight and feedback for management, its ultimate usefulness depends on how widely the results are disseminated. The first step in making information available is to compile, analyze and summarize the results of monitoring. Compilation, analysis and summary should track the same regular schedule along which information is gathered. In this way, the monitoring experience will be fresh in people's minds while they are compiling and analyzing the data, and the results will remain timely.

There are many different options for presenting monitoring results. Both quantitative and qualitative information is presented most effectively with graphs, tables, or pictures rather than simply in a written report. Charts and illustrations can be used to make information accessible even to non-literate audiences. Presenters should always keep in mind the perspectives and background of their audience and tailor presentations accordingly. The guidelines on compiling, analyzing and reporting the results of rural appraisals and other work with communities in sections 3.1.7 and 3.2.7 also apply to monitoring results.

5.3 Implementing evaluation

People evaluate informally all the time, in many different situations. Neighboring farmers evaluate when they compare their relative success in planting different varieties of crops. Managers and residents evaluate when they discuss the impact of an exceptionally wet rainy season on the production of wild fruits commonly harvested from forest trees. In the context of *in situ* management, evaluation is an opportunity for all participants to step back, review past activities and ask broad-reaching questions about their work. Have we addressed management objectives and moved towards our long-term goals for plant resources? Which activities have worked, which have not, and why? In what ways can we improve and reapply efforts to understand and maintain valued plants?

One protocol for a formal evaluation process is presented in Box 5.4. Formal evaluation for *in situ* management can be planned in advance for the midpoint or close of a project, at the end of a growing season, or other logical interval. Evaluation may also be undertaken when problems or conflicts arise related to plant resource management. Formal evaluation, like planning, is usually a group, workshop, or team-based exercise. Generally, a mix of inside participants (protected area managers, local community representatives, project field staff) and outside reviewers (NGO facilitators, rural appraisal experts) provides the best composition for an evaluation team. Insiders are essential participants due to their knowledge of the fieldwork under review, while outsiders can offer alternative perspectives and specific expertise. Also take a look back at the guidelines in section 3.1.1 on preparing for rural appraisal – these are equally relevant for any evaluation effort.

Box 5.4 Steps in an evaluation exercise

1. Identify the reasons for evaluation

The initial step is to make clear to all persons involved why the evaluation is necessary. Potential reasons for an evaluation could be to see if the activities specified in a management plan were actually carried out; to identify whether the activities accomplished met the desired objectives and goals; or to understand why a project is not working well. This step is also a good point to clarify for whose benefit the evaluation is being done, and who will receive the information so generated.

2. Revisit goals, objectives and activities

Next, participants will want to look back at what a management effort intended to do (goals and objectives), and how it was to be done (activities).

3. Formulate evaluation questions

The specific questions the evaluation is designed to answer should be stated at this point. If there are a number of questions to be answered, it may be worth ranking them in order of importance.

4. Review indicators that will answer questions

Indicators for evaluation should correspond closely to the questions that are asked, and may very well be the same indicators measured in monitoring activities. If evaluation indicators differ substantially from those used in monitoring the project or plan, then this suggests that monitoring may not have been directed at the proper subjects.

5. Review available information on indicators

Well designed monitoring exercises should provide much of the information needed for evaluation. However, if monitoring has not been carried out as planned, information may need to be collected as part of the evaluation. In such situations, a premium is often placed on collecting data relatively rapidly. Rural appraisal methods (section 3.1) and other cost-efficient approaches are particularly useful where prior monitoring has not provided an adequate information base.

6. Summarize and present evaluation results

Analysis should directly address the questions that were asked at the evaluation's onset. As with monitoring, evaluation results will be communicated most effectively when presentations are tailored to the audience and use a variety of methods.

Source: [62].

5.4 Tools for monitoring and evaluating plant resources

A wide range of ecological, social, economic and environmental dimensions of useful plants are potential indicators for monitoring and evaluation. A typical project to conserve plant resources may want to measure not only the abundance and productivity of plant populations under management, but also people's attitudes towards the project, their degree of participation, and whether the project is making a difference in their lives. Not surprisingly, the tools and approaches for measuring such indicators and obtaining baseline information are correspondingly diverse. The rural appraisal methods described in Chapter 3, when repeated over time in a careful and systematic fashion, offer ways of qualitatively monitoring a range of social indicators related to effective management of useful plants (Table 5.1). Social indicators can also be assessed through quantitative measures such as censuses and surveys. As previously mentioned, we do not cover these methods here, but they are described in the People and Plants manual *Ethnobotany* [133] and in literature listed in Appendix A. Participatory action research methods (section 3.2.2) offer a format for monitoring and evaluating plant resources collaboratively with local residents. Finally, Chapter 6 describes approaches to monitoring the status of plant resources in agricultural landscapes, in collaboration with local residents.

The remainder of this chapter will focus on quantitative field methods that can be applied in monitoring and evaluating the **ecological** status of plant

Table 5.1 Examples of social indicators that can be monitored and evaluated with rural appraisal methods to judge the success of management for plant resources

Social indicator	Sample means of assessment
Time devoted to gathering/harvesting product	Interviews Daily activity schedules
Distance traveled to gather/harvest product	Mapping Ranking Participant observation
Percentage of monthly income gained from selling plant product	Interviews Matrices Ranking
Degree of community participation in traditional resource management	Interviews Timelines Historical transects Village transects and mapping

resources. We outline a set of broadly applicable field methods for ecological monitoring, but there are plenty of additional approaches to those covered here. Two books in the People and Plants series, Gary Martin's *Ethnobotany* [133] and Tony Cunningham's *People and Wild Plant Use* [58], contain methods for measuring populations and stands of useful plants and for monitoring local harvest of plants for medicine, ceremony, commercial harvest and home subsistence. Additional references that review conservation tools for monitoring and evaluation are listed in Appendix A.

5.4.1 Monitoring and evaluating ecological status: levels of intensity

Monitoring a species' ecological status at a level that is thorough, yet also efficient, is a big challenge for managers and conservation professionals. Some researchers spend their entire careers studying a single plant or animal species, detailing how it survives, reproduces and interacts with other organisms and its environment. This is a noble pursuit, to be sure, but not one that most conservation programs have the luxury of undertaking. The challenges that land managers face is to gather sufficient information on ecological and life history factors to yield insight for managing and monitoring a species or population of concern.

Ecologists Eric Menges and Doria Gordon [140] have addressed this issue in their work with rare plants in the southeast United States. They advocate a system of ecological monitoring and evaluation based on three levels of intensity (Table 5.2). On the first and least intensive level, locations of habitat types and species occurrences are mapped and re-visited at intervals to note their persistence or disappearance. No attempt is made to quantify precisely the size or structure of populations, although visual assessment of abundance (section 4.1.2) is used on occasion. The spatial dimensions of stands of priority species are sometimes mapped and repeatedly inventoried as well. The field methods used at this

Table 5.2 Three levels of ecological monitoring intensity for plant resources (adapted from Menges and Gordon, 1996)

Level of investigation	Monitoring aims	Field methods	Selected measurements
Occurrence and distribution of populations	Measure trends across populations or species at a site Hypothesize trends in population size	Presence/absence surveys Mapping species occurrences Plotting boundaries of populations and/or stands Ocular assessment Habitat quality assessment	Area occupied by stand Number of populations at site (these are best viewed as semi-quantitative or descriptive)
Population size and condition	Measure trends within populations Hypothesize mechanisms for observed trends	Plot sampling (e.g. quadrats) Transect sampling (e.g. point-quarter) Cover assessment Logarithmic plot sampling	Abundance Density Percentage cover Size class structure Reproductive status (these last two may be semi-quantitative)
Population demographics	Understand mechanisms that influence population behavior Predict future population trends	Mark and follow individuals over time Measure density and abundance of individuals in size, stage, or age classes (can use plots or transects) Measure seed production, germination, establishment Measure clonal reproduction Life history studies (e.g. pollinators, dispersers, herbivory)	Mortality rates Reproductive rates Size/stage distributions Recruitment rates Rate of population growth Rate of population turnover Pollination rates Dispersal frequencies or distances Rates of herbivory

most general level are essentially those discussed in Chapter 4 under inventory approaches for species and habitat types.

At the second level of effort, habitat composition and species abundance and density are quantified using plots, transects and other field survey methods. Where the relative age of individuals can be readily assessed (as by measuring trunk diameters of forest trees), this information may be quantified along with density and abundance. This level of monitoring is often appropriate for species important in local economies, habitats rich in phytogenetic resources, or other management priorities.

At the third and most intensive level of monitoring, the demographic structure of species is assessed by marking and censusing numbers of individuals in different age or stage classes, and assessing reproductive rates, survivorship and other demographic variables. This level of monitoring also involves understanding detailed aspects of the life history of a species. It is likely to be applied only to the highest priority species, such as rare wild relatives of crops or highly valued ethnobotanical resources.

As we present methods for monitoring and gathering baseline ecological information at these different levels, we reiterate that such investigations lend themselves readily to participation and collaboration with local residents. Indeed in many cases, a collaborative approach will produce more timely and relevant results.

5.4.2 Density and abundance (level 2): plot sampling

For species identified as protection priorities, it is desirable to have baseline information on abundance, density and reproductive status. If a plant population or its habitat is small and localized, it may be possible to survey the entire population by counting and measuring all individuals. For most species, however, such exhaustive evaluation is an unrealistic task. As a result, baseline information on plant resources is usually collected by taking a random sample of a population or community and extrapolating from the sample results to the entire site or target population. One of the most common means of sampling is to lay out a series of **plots** or **belt transects** of a standard area, in which individuals are tallied and measured or otherwise quantified. (Belt transects are essentially long, skinny plots, such as 10 m × 1000 m. They have the advantage of being able to sample efficiently a variety of habitats and/or microsites, because they are long enough to run across local environmental gradients – say, from ridgetop to stream bottom, or from floodplain to *terra firme* forest.) Plots are also commonly used to sample the species diversity and composition of habitats and plant communities at a site. To conduct plot sampling effectively, you need to consider the following questions.

(a) How should you arrange plots at a site?

Selecting and maintaining a consistent arrangement of plots is the first step in obtaining high-quality plot data. Foresters and ecologists have developed a variety of methods for arranging plots. Following the work of Charles Peters [166] on inventorying non-timber forest resources, we will focus on three methods here:

- **systematic** or regular sampling;
- **random** sampling;
- **stratified-random** sampling.

In **systematic sampling**, plots are laid out across a stand or site in a regular fashion at fixed intervals. For instance, in a survey of the tree composition of a 100 ha forest stand on the Caribbean island of Dominica, plots were established at 150 m intervals along parallel transect lines that followed a northeast compass bearing (Figure 5.2).

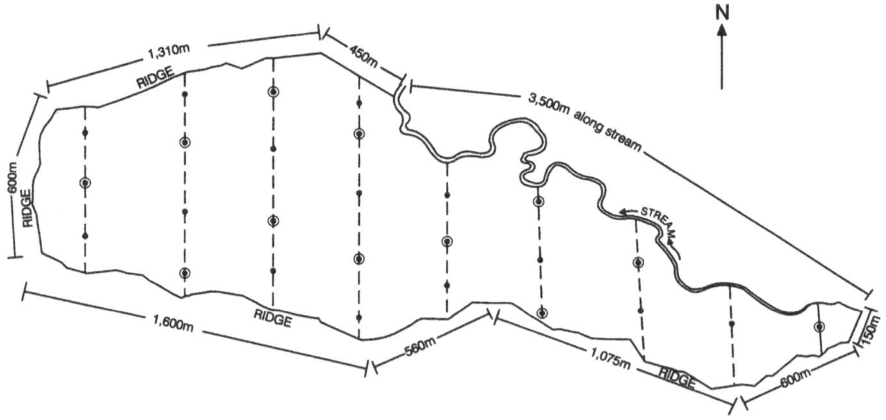

Figure 5.2 Plot sampling design for assessing composition and diversity of a forest stand in Bois Diable, Dominica. [[encircled dot]], plots of fixed area (0.2 ha); [[dot]], prism-surveyed plots. Plots were laid out at approximately 150 m intervals along parallel transect lines 450 m apart and 150 m from each end of the plot. (Adapted from Tuxill, 1991.)

The advantages of a regular sampling pattern are that it estimates population abundance and density well; it can be done without knowing the total area of the site in advance; it facilitates mapping and qualitative data collection across the entire site; and it is efficient to implement because of the relative ease of locating and travelling between plots. The drawbacks of a systematic pattern are that it is not fully random (while the initial plot is located randomly, all others are placed at fixed intervals from that first plot); therefore certain statistical analyses are not possible, such as estimating the degree of sample error. Systematic samples at different sites cannot be compared statistically either. These are not fatal flaws – systematic sampling will still provide good estimates of the density and abundance of individuals – but they do limit the analyses that are possible with systematic sampling.

In **random sampling** (Figure 5.3a), each plot is located independent of all

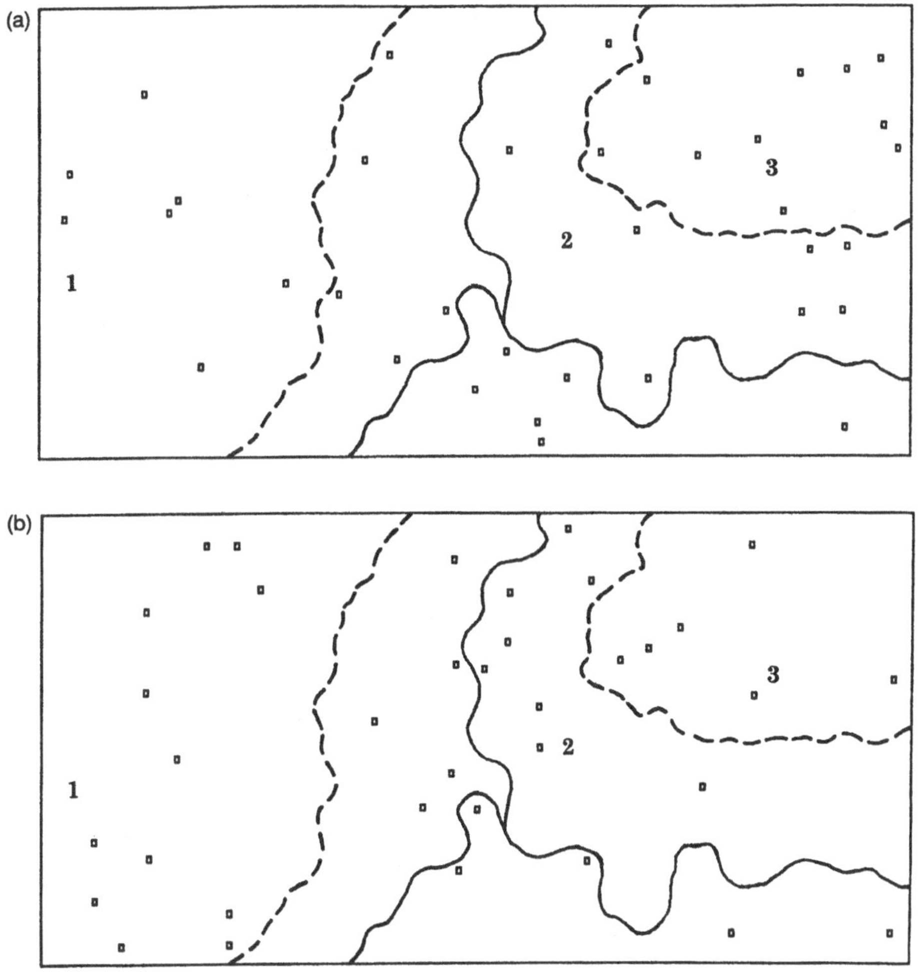

Figure 5.3 These two maps illustrate the difference between random (a) and stratified random (b) sampling designs. Both maps cover three habitats, separated by dashed lines. Habitats 1 and 3 are upland areas, each having a different soil type; habitat 2 is a riverine floodplain. In both maps 36 sampling plots (small squares) were located by placing a coordinate grid over the map and selecting coordinates for each square at random. In map (a) the squares were placed without any regard for the underlying habitat zones. The random pattern that resulted oversamples habitat 3 (ten squares) and undersamples habitat 1 (seven squares), based on the relative area of each habitat. In map (b) each habitat was assigned a quota of plots according to the area it covered – thus habitat 3, which covers one-sixth of the map, recives six plots. Note that within each habitat of map (b) the plots are still located randomly using the same grid that was used to locate the plots in map (a); they are simply stratified between habitats. (based on Peters, 1995.)

other plots. One way to randomize plot location is to lay a coordinate grid atop a map of the site to be sampled and then select coordinates at random (for instance, by drawing slips of paper labelled with coordinates out of a hat), with each pair of coordinates corresponding to a plot location on the map [166]. The distances and compass bearing needed to arrive at each plot can then be calculated directly from the map to locate plots in the field. The strength of random plot arrangement is that it supports full statistical analysis, including calculation of the degree of error involved in the sampling and full comparison between sites. The drawbacks to random arrangement are that it can leave substantial areas of a site unsampled (due to random clumping of plots); it does not facilitate site mapping as well as systematic sampling does; and it is often more tedious to implement because of the greater time required to locate and travel between plots.

To reduce these limitations, you can opt for the third arrangement: **stratified-random sampling** (Figure 5.3b). In this approach plots are located randomly, but only after first assigning, or **stratifying**, a certain quota of plots to previously noted habitats or microenvironment types – such as may be defined by soil, moisture availability, altitude, slope, aspect, stand age, or other features that influence vegetation. The quota of plots received by a given habitat or microenvironment should correspond to the percentage of area which that habitat occupies at the site being sampled [166]. Consultations with local residents can help managers to identify appropriate divisions for stratifying plots at a site. The same randomized-coordinate method can then be used to situate plots, with the simple adjustment that once the specified quota of plots has been drawn for a habitat, any subsequent coordinates drawn corresponding to that habitat are ignored until all habitat types or microenvironments have their appropriate quota of plots.

(b) How large should you make the plots? How many replicate plots are necessary? Answering these two related questions properly depends upon several factors: the size of the plants being studied (density tends to be inversely correlated with size), the pattern of distribution exhibited by individuals (widely scattered vs. tightly clumped) and the degree of statistical accuracy desired. Table 5.3 presents some plot sizes and replicate numbers suggested by Poffenberger and colleagues [175] as general benchmarks for assessing trees, shrubs and herbs that will generate good statistical results. In addition, Peters [166] recommends for forest inventories that the total area sampled in plots should be at least 3–5% of the total area of the site or stand that is being studied. If plots are being used to sample the species diversity and composition of a site, then a graph of species tallied against area sampled can also be useful (Figure 5.4). When this plotted species–area curve flattens out, indicating that few new species are being recorded for each new area sampled, it is a good sign that your sample size has captured most species and is adequate for the habitat you are sampling.

Table 5.3 Suggested quadrat sizes for monitoring plant populations and assessing vegetation (adapted from Poffenberger *et al.*, 1992, by kind permission of Society for Promotion of Wastelands Development, New Delhi, India)

Vegetation type	Quadrat dimensions (m)	Number of replicate quadrats	Comments on vegetation assessment
Trees	50 m × 20 m	5 to 8	If topography, soil type or vegetation patterns are highly variable, increase number of quadrat replicates
Shrubs and tree seedlings	5 m × 5 m	10 to 16	For each tree quadrat, include two shrub/tree seedling quadrats
Herbs and grasses	5 m × 5 m	20 to 32	For each shrub quadrat, include two herb/grass quadrats

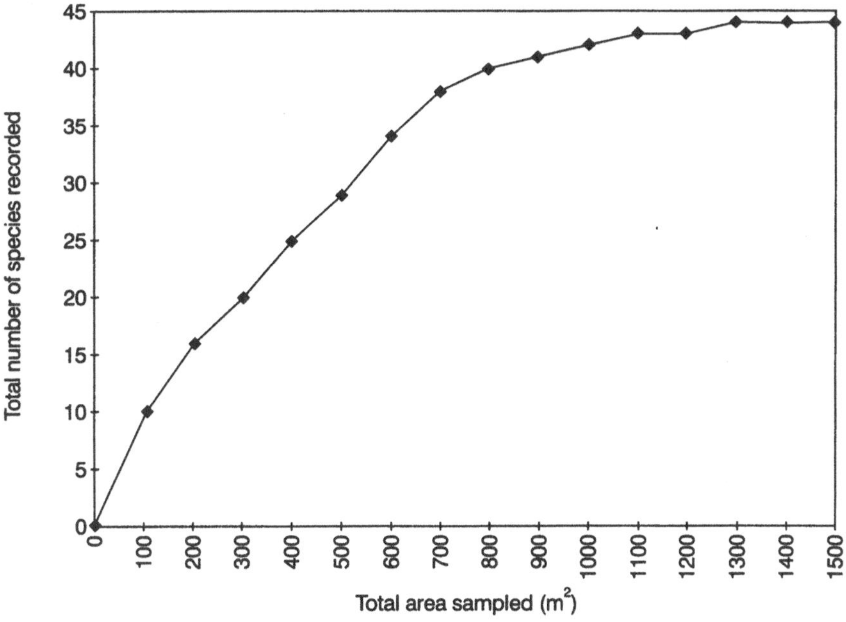

Figure 5.4 A sample species–area curve showing how the total number of tree and shrub species tallied in a 150 × 10 meter belt transect gradually increases as each 10 × 10 meter section of the transect is sampled. Note that by the 90 meter mark (900 square meters) most of the species have already been encountered at least once, and the rest of the transect adds relatively few species. From 900 to 1500 meters the curve is virtually flat or horizontal, indicating that this plot is has probably captured the majority of tree and shrub species present in this habitat.

(c) What will you need to lay out plots?

The equipment required includes:

- a measuring tape for pacing off the plot dimensions;
- stakes, paint, flagging tape, or string to mark plot corners and boundaries;
- a directional compass for use in orienting the plot and squaring off corners.

Survey plots for plants are most commonly rectangular in shape, but square quadrats or circular plots are used as well. Usually, one corner of the plot is marked first and then the sides are measured out with tapes. Often, there may be some question about whether or not to count plants that fall along the plot border. Any plants that have more than 50% of their trunk (or other measure of size being used, such as canopy) inside the plot should definitely be measured. Individuals that sit precisely on the plot boundary can be alternately counted and excluded – the first one counted, the second one disregarded, the third counted and so on [166].

(d) What information should you record?

As a minimum, plot surveys usually record the names (both scientific and local) of species present and count or estimate numbers present of individuals of priority species (or of all species, for a habitat inventory). If you are not well versed in the local flora or do not have species descriptions to work from, you will need to collect vouchers of the species you encounter while sampling and identify them afterwards in an herbarium. Local residents with extensive botanical knowledge (in most rural areas there are at least one or two such individuals per village) can often identify by local name the bulk of the flora sampled during plot inventories. Folk botanical experts are skilled at distinguishing between species based on vegetative characteristics, such as the color, texture and smell of the inner bark of similar looking trees. Be aware, however, that folk taxonomies do not always correspond precisely to scientific taxonomies. Sometimes several similar species are lumped under a single local name, or a single variable species may have several local names [84, 108]. Thus when sampling is likely to involve a large suite of little-known species (as in many tropical forests), collaboration with a professional botanist is also highly recommended.

You should clearly define beforehand what will be tallied as an individual plant. Identifying a true genetic individual can be difficult for certain plants that spread vegetatively via tillers or rhizomes. Simple visual inspection often cannot reveal whether each clump of grass or tree stem in a cluster is the product of a single seed (in which case each clump or stem is a **genet**) or whether each cluster is a single individual that has undergone vegetative propagation (making each clump or stem a **ramet**).

The only clear way around this problem is to decide at a survey's onset whether genets or ramets will be counted and then count consistently throughout the survey. Such information should be noted clearly on the survey form so that there is no ambiguity when the time comes to tally and analyze survey results. Remember as well that samplers for subsequent years should be trained to use the same units for comparison.

It is also common in plot samples to assign individuals to size classes or stage classes (see section 5.4.5), and to note flowering or fruiting status. This information can often be collected relatively quickly and provides an initial picture of population structure. Size classes for forest trees are usually based on measuring trunk diameters (Box 5.5). However, many plot samples only measure diameters above a minimum diameter limit, such as 10 cm, because numbers of individual trees tend to increase exponentially with decreasing size. Qualitative information about the appearance of a plot (soil type, slope, aspect, moisture availability, evidence of past human management) is important to record as well. These can offer clues about a species' ecology, often indicating whether it is dependent upon a particular microenvironment or ephemeral successional stage. Obtaining local participants' impressions of a plot is especially important, because they are likely to be more aware than outside ecologists of historical land uses or similar factors that may have influenced a particular site [133].

5.4.3 Measures of cover (level 2)

In addition to density and abundance of individuals, a third attribute that can provide useful information on a plant population is **cover**, or the amount of ground area taken up by a plant when viewed from above. Cover is calculated as the area occupied by individuals of a plant species, divided by the total area of habitat sampled – commonly expressed as m^2/ha. **Foliar** or **canopy** cover is a measure of the area covered by the foliage of a plant; it is measured as the diameter of a plant's canopy across its densest portion, approximated to a circle for calculating area [35]. For trees, cover is usually measured as the **basal area** occupied by a tree's trunk, measured as diameter at breast height (Figure 5.5).

When measuring individual plants is impractical or exceedingly laborious, such as with dense herb or shrub vegetation, then researchers commonly estimate cover as a percentage value. Percentage cover can also indicate the sparseness of a forest canopy or ground vegetation and thus be used to monitor overall habitat quality. For forest canopy cover, Poffenberger and colleagues [175] describe a rapid assessment method that relies upon a transect approach and visual estimation. Starting at a random point in a forest stand, the observer walks a compass line and stops every 5 m to assess the tree canopy directly overhead. The amount of canopy closure is then placed into one of four classes (Figure 5.6) based on the amount of overlap between tree crowns. A total of 30 observation points (equivalent to a transect of 150 m) is recommended as an adequate sample size, from which can be calculated mean score and percentage of canopy closure.

To quantify the relative cover for herbs or shrubs, the percentage canopy cover of each plant species can be visually estimated within plots or belt

Box 5.5 Taking diameter measurements of forest trees

Tree diameter measurements are usually measured around the trunk at a 'breast height' of 1.3 m off the ground, hence the common abbreviation 'dbh'. Dbh figures can be calculated from circumference readings taken with standard measuring tapes (don't forget to divide by π), but special measuring tapes are available through forestry supply companies that are calibrated to give diameter measurements directly. For trees that have buttresses or other trunk irregularities at dbh height, the point of measurement is moved up or down the tree to obtain a reading that best reflects the actual size of the tree trunk (Figure 5.5).

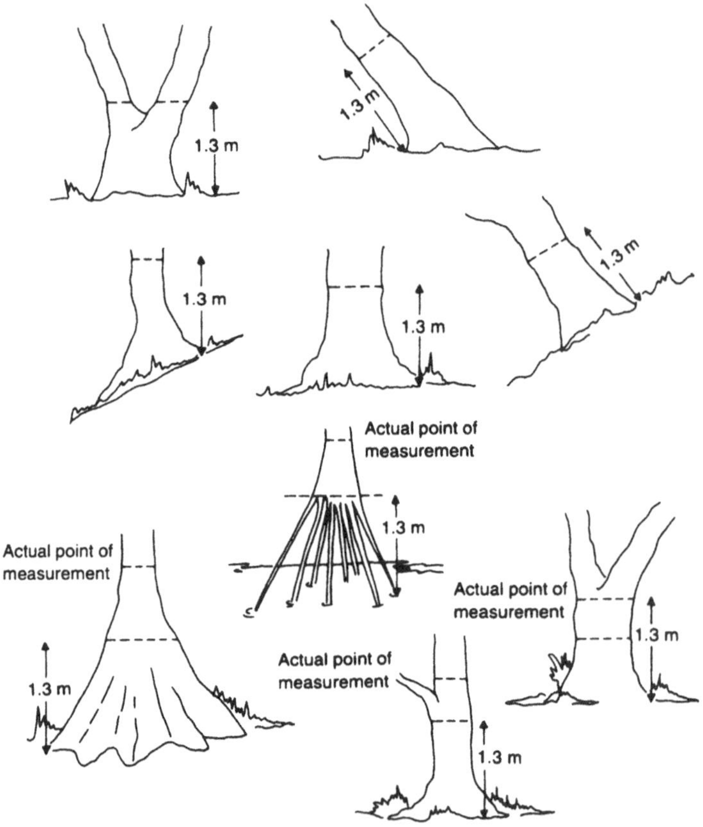

Figure 5.5 Techniques for measuring the diameter at breast height (dbh) of trees under various circumstances. Center: normal dbh measurement of a straight, single-trunked tree on level ground. Top: how to account for split trunks, leaning trees and unlevel ground. Bottom: how the actual point of measurement should be moved up or down to avoid trunk abnormalities such as buttresses, stilt roots, branches and forks. (From Martin, 1995.)

Figure 5.6 Forest canopy cover assessment: a rapid qualitative method. Classification and percentage of closure cover: 0 = no canopy cover (0–20%); 1 = adjacent crowns meet (40%); 2 = crowns overlap, but still allowing light through (80%); 3 = sky not visible (100%). (From Poffenberger *et al.*, 1992, reprinted by kind permission of Society for Promotion of Wastelands Development.)

transects and assigned to one of a range of percentage classes along a sliding scale: 0–5%, 6–25%, 26–50%, 51–75%, 76–95%, 96–100%.

When averaging cover percentages across several plots (Table 5.4), the **midpoints** of the classes are used. For the above classes, the midpoints are: 2.5%, 15.5%, 38%, 63%, 85.5%, 97.5%.

Table 5.4 Sample measurements of ground-layer vegetation under a semi-arid thornscrub canopy. The vegetation was sampled with three square plots each 5 × 5 meters, where each species indicated was assigned a cover class from the scale represented in the text.

Species	(Three plots total) Cover score per plot 1	2	3	Mean cover value (calculated using midpoints of percentage cover classes)
Desmodium	1	2	2	(2.5 + 15.5 + 15.5)/3 = 10.8%
Gossypium	5	3	2	(85.5 + 38.0 + 15.5)/3 = 46.3%
Bouteloua	1	1	4	(2.5 + 2.5 + 63.0)/3 = 22.7%
Cucurbita	0	1	1	(0 + 2.5 + 2.5)/3 = 1.7%
Mimosa	2	3	1	(15.5 + 38.0 + 2.5)/3 = 18.5%

The plot size guidelines for shrubs and herbs in Table 5.3 are also suitable for estimating cover. Remember that since this method relies upon visual assessment, and also groups the cover estimates, the values for cover obtained with this scale cannot avoid a certain degree of bias and should be considered as general estimates.

Conservation professionals working in semi-arid or arid environments may wish to consider a vegetation assessment method that measures both density and cover, developed by ecologist Joseph McAuliffe [130]. This approach, the logarithmic plot method, is designed for assessing vegetation where individual plants have visually distinct, easily estimated canopies (Box 5.6).

Box 5.6 The logarithmic plot method for arid-lands vegetation

Because plants in arid and semi-arid habitats tend to be widely scattered and have diverse architecture, they do not lend themselves well to sampling for density or cover via standard quadrats. The log-survey method avoids these problems by measuring species abundance and cover in plots on a logarithmic scale. Here's how it works [130].

1. Equipment needed:
   - metric measuring tape, 25 m in length;
   - 1 m stick, to measure plant canopy diameters (one for each person);
   - pencil and notebook for recording data;
   - (optional) calculator, for data analysis;
   - (optional) compass, if laying out transect lines.
2. At the site chosen for monitoring, select points for plots, following the guidelines in section 5.4.2. At each point, lay out a circular plot of either 9.03 m radius (256 m^2) or, if the vegetation is particularly sparse, 12.77 m radius (512 m^2). These plot dimensions probably seem unusually specific at first glance, but they are related to the logarithmic evaluation.
3, Tally abundance of individuals for all plant species in the plot, assigning each species to an abundance interval (Figure 5.7, lower right box). As illustrated in the sample data sheet, each abundance interval is twice as large as the previous one, and the midpoints of each interval form a geometric series: 0, 1, 2, 4, 8, 16, 32, and so on. This ranking system does two things: it preserves small distinctions between species that occur at low abundances; yet it also provides for rapid assessment of species that occur in larger numbers, because the species are only assigned to interval classes, not counted precisely. Very abundant species are best estimated by tallying the individuals in a one-fourth or one-eighth wedge of the plot circle, and then scaling up accordingly.

Logarithmic Plot Data Sheet

SPECIES	Abundance: No. of indivs.	Abundance: Log class	Cover: Average diameter (m.)	Cover: Log class
Boutelova curtipendula	11	3	0.40	-3
Janusia gracilis	43	5	0.52	-2
Calliandra eriophylla	60	6	0.67	-2
Boutelova chondrosoides	23	4	0.15	-6
Heteropogon contortus	196	8	0.30	-4
Bothriochloa barbinodis	7	3	0.20	-5
Fouquieria splendens	4	2	1.05	0
Celtis pallida	3	2	1.30	0
Aristida sp.	12	4	0.20	-5
Sporobolus sp.	5	2	0.25	-4
Opuntia violacea	1	0	0.50	-2
Gossypium thurberi	9	3	0.33	-4
Prosopis velutina	2	1	3.50	3
Mimosa dysocarpa	4	2	0.40	-3
Alloysia wrightii	1	0	0.50	-2
Cercidium floridum	1	0	0.20	-5
Acacia greggii	1	0	3.00	3
Mimosa biuncifera	1	0	0.80	-1
Mammilaria grahamii	1	0	0.05	-9

Plot ID #: 16

Location:
ROCK CORRAL CANYON,
TUMACACORI Mts.,
ARIZONA
UTM LATITUDE 49397
UTM LONGITUDE 3492747

Date Surveyed:
16 FEB 1994

Observers:
G.P. NABHAN
J. TUXILL
J. DONOVAN

Notes:
SITE AT 1,020 m. ELEVATION,
ON 20 DEGREE SLOPE,
SUBSTRATE OF 20-40 cm.
DIAMETER COBBLES,
UPLAND SEMIARID SCRUB

Converting cover to log cover class

mean cover/indiv. (m^2)	area interval (m^2)	diameter interval (m)	log cover class
0.002	< 0.0032	< 0.06	-9
0.004	0.0032-0.0064	0.06-0.09	-8
0.008	0.0064-0.011	0.09-0.12	-7
0.016	0.011-0.023	0.12-0.17	-6
0.031	0.023-0.045	0.17-0.25	-5
0.063	0.045-0.10	0.25-0.35	-4
0.125	0.10-0.19	0.35-0.49	-3
0.25	0.19-0.38	0.49-0.69	-2
0.5	0.38-0.75	0.69-0.98	-1
1.0	0.75-1.5	0.98-1.4	0
2.0	1.5-3.0	1.4-2.0	1
4.0	3.0-6.0	2.0-2.8	2
8.0	6.0-12.0	2.8-3.9	3
16.0	12.0-24.0	3.9-5.5	4
32.0	24.0-48.0	5.5-7.8	5

Coverting abundance to log abundance class

interval	midpoint	log class
0	0	-1
1	1	0
2	2	1
3-5	4	2
6-11	8	3
12-23	16	4
24-47	32	5
48-95	64	6
96-191	128	7
192-383	256	8
384-767	512	9
768-1,535	1,024	10
1,536-3,071	2,048	11
3,072-6,143	4,096	12

Figure 5.7 Sample data sheet for logarithmic plot assessment, with conversion tables for abundance and cover values.

4. Record the logarithmic abundance class for each species; this class is the logarithm-base 2 of the midpoint of the abundance interval, and can be read directly off the conversion table in the lower right box of Figure 5.7. This log abundance class value will come in handy in step 7 below.

5. Tally cover for each species in the plot. This is done by assigning each species to an average individual canopy diameter interval (Figure 5.7, lower left box). As with abundance, these intervals are arranged on a geometric scale, with the midpoint of each interval twice as large as the previous one. Canopy diameters rather than canopy areas are used because it is far easier to estimate and measure a linear diameter dimension.

6. Record the logarithmic cover class for each species, this also being the logarithm-base 2 of the cover interval midpoints (Figure 5.7, lower left).

7. To determine the total cover per species per plot, add together the log cover and abundance class values. This will give the log cover class value for total cover per species, and can be readily translated back into an actual m^2 cover estimate using the lower left table in Figure 5.7. For instance, the total cover class value for *Calliandra eriophylla*, a shrubby legume, is $6 + (-2) = 4$. Reading from the final column of the tables, the log cover class of 4 translates back into a total cover estimate of 16 m^2. This estimate is easily translated into a percentage cover value by dividing the total cover estimate by the total area of the plot. The plot sampled in Figure 5.7 was 9.03 m in diameter, or 256 m^2, so *Calliandra eriophylla* occupies 16/256, or 6.25%, of the plot.

The mathematics of the log-survey method may seem complicated at first glance, but once you become familiar with the calculations, the efficiency it offers for tallying data is soon apparent. Since the log-survey method provides both abundance and cover values, it enables fieldworkers in arid and semi-arid terrain to gather density, cover and diversity information in as little as 20 minutes per plot. The main limitation is that log-cover surveys remain efficient and accurate only when the canopies of individual plants are distinct and cleanly measured, as in open shrub communities. Closed-canopy forests and grasslands will probably remain better suited to other cover measures.

5.4.4 Successional dynamics (level 2)

Successional dynamics are also referred to as **disturbance regimes** or **patch dynamics**. Any landscape is a patchwork of habitats at differing stages of recovery from periodic disturbance. Some disturbances occur on a small or **fine-grained** scale, such as when an ageing tree weakens and crashes to the forest floor, leaving a small sunlit gap in the forest canopy. Others, such as a typhoon that strikes a Pacific island chain, or the seasonal floods that inundate large areas of the lower Amazon floodplain in Brazil, leave a much larger imprint upon the landscape when they occur. Such episodic disturbances, whether large or small, serve to maintain a variety of vegetation associations and microhabitats within a landscape, enhancing both structural complexity and species diversity.

For effective *in situ* management, it is important to understand how natural disturbance regimes help to create or maintain habitats for species that are management priorities. If several rare plants are found to flourish in a fire-dominated chaparral community, it may be necessary to conduct periodic prescribed burns. Alternatively, if an economically important species requires an advanced successional stage or **old-growth** community, then management activities will have to focus on limiting disturbance and providing enough old-growth habitat to support viable populations of the resource species.

Quantifying successional patterns can pose quite a challenge. Because succession is a gradual ongoing process, it cannot be observed directly at any one point in time. The most precise and unambiguous way to measure successional dynamics is to: establish permanent vegetation plots (marked at the corners); census all plants in the plots (noting species and size for each); and re-census each plot at two- to four-year intervals to quantify changes in structure and composition of the vegetation [18].

There are several drawbacks to using permanent vegetation plots alone for assessing succession. To start with, there will be a lag time of at least several years before successional measurements become available. With long-lived plants, including most forest trees and semi-arid shrubs, it can take decades for successional patterns to become evident in permanent plots. Managers charged with conserving plant populations that are declining rapidly or facing immediate threats usually do not have the luxury of waiting several years for information on successional dynamics. Also, over time it may be difficult to relocate plots or maintain the institutional commitment of organizations and agencies to monitor plots at regular intervals.

Several short-cut methods, when applied carefully, can give an accurate picture of successional patterns in a particular landscape or region. Perhaps the most important step is to draw on the knowledge of local residents. Village elders and others who have spent a lifetime observing nearby landscapes often possess precise knowledge on the location and timing of significant past disturbance events, such as fires, blowdowns from windstorms and hurricanes, and vegetation clearance for agricultural plots. Interviews, mapping and other participatory appraisal activities with residents can indicate where best to locate plots for sampling vegetation representative of both a specific age and specific disturbance event.

Collaborating with local land-use experts can enable managers to quantify vegetation composition and structure (using plots or transects) on sites that have regenerated from past disturbance at an established point in time. By comparing sites of known but different ages, it is possible to infer patterns of disturbance and succession prevalent in a landscape. As with laying out replicate plots, it is also important to consider microenvironmental factors – slope, aspect, elevation, soil type, moisture availability – that influence vegetation and create differences between plots.

Used together, sampling plots of different ages can yield valuable management information on successional dynamics. For instance, botanist Bruce Benz and colleagues [23] from the University of Guadalajara in Mexico have investigated the successional vegetation required by *Zea diploperennis*, a wild perennial relative of corn, endemic to a single mountain range (Box 5.7).

Box 5.7 Zea diploperennis and habitat succession

When managers and researchers at the Sierra de Manantlan Biosphere Reserve in Jalisco, Mexico, set out to protect their flagship species, *Zea diploperennis* (known locally as *milpilla*), they soon began gathering baseline information on the species' habitat requirements. Initial qualitative surveys revealed that all known *milpilla* populations are near highland farming villages, and that the plants invariably occur in sunlit clearings surrounded by pines, oaks and broadleaf cloud forest. Most of the sites favored by *Z. diploperennis* appeared to be created by small-scale clearance for maize cultivation. Indeed, in some cases the plants were growing in actively cultivated fields.

To clarify the successional dynamics of these sites, researchers quantified the abundance of *milpilla* and associated vegetation in fields and fallows (abandoned fields) of different ages. They selected two disjunct populations for sampling, and stratified their samples among three stands in each population. By consulting with local farmers, they identified *milpilla* stands of known successional age, from presently cultivated (age 0) to 1, 2, 5, 10 and 15 years post-cultivation.

At each stand, they took randomly placed vegetation samples using quadrat plots 1 m by 1 m square. In each quadrat, all individual stems of *Z. diploperennis* and their total percentage cover were tallied, while all other vascular plant taxa in the quadrats were recorded on a presence/absence basis. The physical appearance of each stand was also noted, including exposure, soil depth and compaction, slope, and amount of bare ground. Farmer interviews yielded additional information, such as whether herbicides or fertilizers had been used during cultivation, whether the site had been grazed, and what the original vegetation had been when the site was first cleared.

The results demonstrated that *Z. diploperennis* is the dominant plant, measured in terms of cover, at all disturbed sites. In addition, the species increased in cover and stem abundance over time, with both figures highest in the 15-year-old plot. Apart from site age, no other physical or historical features of the sites correlated with these trends. The majority of this growth appeared to be due to an increase in ramets from established plants rather than new genetic individuals. Moreover, the 15-year-old plot revealed the first incursion of young woody trees that could eventually shade out *milpilla* and other herbaceous plants. These findings suggest that while *milpilla* is adept at colonizing and dominating forest openings for up to 15 years, its long-term persistence in the Manantlan Biosphere Reserve will depend upon regular small-scale forest openings like those produced by shifting agriculture [23].

Certain successional trends can also be discerned from the stage or size-class distribution of the dominant plant species at a site. Foresters will recognize this method immediately, as it is widely used to quantify forest vegetation and tree regeneration. By comparing the species composition and relative abundance of the tree overstorey, pole tree, sapling and seedling stages, recent successional trends at a site can be inferred. For instance, a situation where the dominant overstorey species are not abundant in the seedling or sapling stage suggests that the overstorey is dependent upon some form of disturbance for establishment, such as fire that opens up a closed grove to high light levels and promotes seed germination. In the absence of that disturbance, the overstorey vegetation will eventually die and be replaced by plants that can germinate and thrive under the canopy.

5.4.5 Quantifying demographics (level 3)

While assessing the demographic structure of a population is usually the most laborious form of ecological monitoring, it also provides information not available through other sampling methods. Demographic structure is particularly important for assessing regeneration dynamics of a plant species, as it provides information that cannot always be gleaned from figures on total population size (Figure 5.8).

What information is needed for demographic monitoring? Ideally we would like to know how many individuals are in a population, how these individuals are distributed across the various age classes within the population and how these numbers change over time. This information can be revealed by sampling the population, marking individual plants, evaluating their relative age, measuring their reproductive success in terms of seeds and seedlings produced, and following them over time.

Demographic assessments generally tend to measure the age of plants in terms of size or life stage classes. **Size classes** are most commonly used with closed-canopy forest trees, which increase steadily in girth as they age, and are based on trunk diameters (see Box 5.5). **Stage classes** are often used with shrubs, cacti, palms and other plants for which stem size is difficult to measure or, due to the physiology of the plant, not a good reflection of the plant's age. This involves quantifying numbers of individuals in a range of life stages:

- seeds
- seedlings
- juveniles
- reproductive-age individuals
- senescent (post-reproductive) individuals
- dead plants.

Although assessing fruit and seed productivity is a particularly time-consuming activity, it will improve the accuracy and value of demographic monitoring. For instance, if seed productivity is high yet few seedlings are recorded, then a

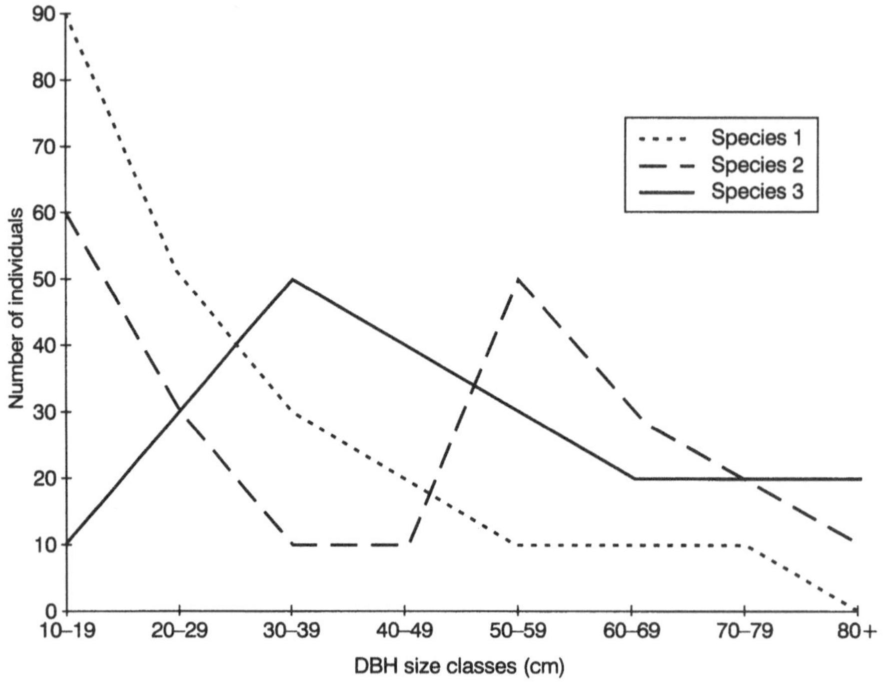

Figure 5.8 Graph showing how demographic assessment can provide information not evident when only total abundance is measured. All three tree species have the same population size (220) but exhibit very different distributions of individuals within size classes. Species 1 shows strong recent reproduction and establishment, and is evenly distributed throughout all but the largest size classes. The population of Species 2 is dominated by individuals in two size classes, indicating that it may require occasional disturbance events (such as fire or a hurricane) to perpetuate itself. Species 3 is dominant in the largest size classes and all but one middle size class, but has not been reproducing or establishing well in recent years. Of these three, Species 3 appears most likely to need monitoring – and perhaps conservation efforts, if it continues to decline.

plant may be producing primarily non-viable seeds, there may be high predation, dispersal may not be occurring, or environmental conditions may not be suitable for germination. It can be useful as well to sample the number of seeds produced per fruit, for comparisons with seed levels in other individuals and populations. Seed viability and seedling vigor can be assessed through germination trials and seedling transplants. A straightforward test for viability can also be done with the chemical tetrazolium (Appendix A). In addition, sometimes just the weight or appearance of a seed can be a reliable indication of its viability.

As with other kinds of ecological monitoring, demographic studies can be implemented in collaboration with local residents. One example where researchers and local residents have worked together on demographic monitoring comes from northwest Ecuador, where local and international conservation groups are promoting ecologically sound management of tagua palm groves native to the region (Box 5.8). As part of these efforts, local tagua nut collectors were trained in population monitoring, worked alongside a biologist to collect demographic measurements, and have continued to work as para-biologists, collecting data after the departure of the researcher from the study area.

Box 5.8 Local demographic monitoring of a tagua palm in Ecuador

Phytelephas aequatorialis is a palm native to the northwest Pacific slope of the Andes. It has long been valued for its hard ivory-like seed called tagua, which can be carved into buttons, figurines and jewelry. Within its range, *P. aequatorialis* is most common along riverbanks and other early successional habitats. It is also a naturally regenerating component of farmers' agroforestry plots, along with cacao and other crops.

Since 1989, the group Conservation International and an Ecuadorian organization, CIDESA, have sponsored a project in northwest Ecuador to improve markets for tagua nuts and promote ecologically sound management of the resource. During 1993, Julie Velásquez Runk [235] worked with the project to design and implement a monitoring system for tagua stands at La Comuna Río Santiago-Cayapas that would train local residents to assess the ecological sustainability of tagua harvest and management techniques. Part of this work examined the demographic structure of tagua stands under various local management methods (Figure 5.9).

Data collection involved establishing 20 × 30 m rectangular plots; tallying numbers of juvenile, subadult and adult palms present; and noting or calculating adult palm density, distribution, sex, stem height, light exposure, foliar coverage, phenological status, total number of live and dead leaves, and presence of inflorescences and/or infructescences. Seed productivity and viability were also assessed. Data were collected every three months. By mapping and tagging individual palms, combined with painting the leaves of juvenile palms and the stalks of inflorescences and infructescences, researchers measured recruitment into the population and compared it with seed productivity.

Velasquez-Runk and Comuna residents worked together at all stages of the fieldwork. Local tagua collectors identified suitable locations for study plots, and measured and weighed tagua seeds as they were collected. Several tagua collectors, local students and volunteers also received training as parabiologists, during which they were introduced to the research rationale and were trained in all data collection methods. For six months the parabiologists accompanied the researcher in surveys of tagua plots, and then assumed primary responsibility for data collection.

Together the researcher and parabiologists collected over two years'

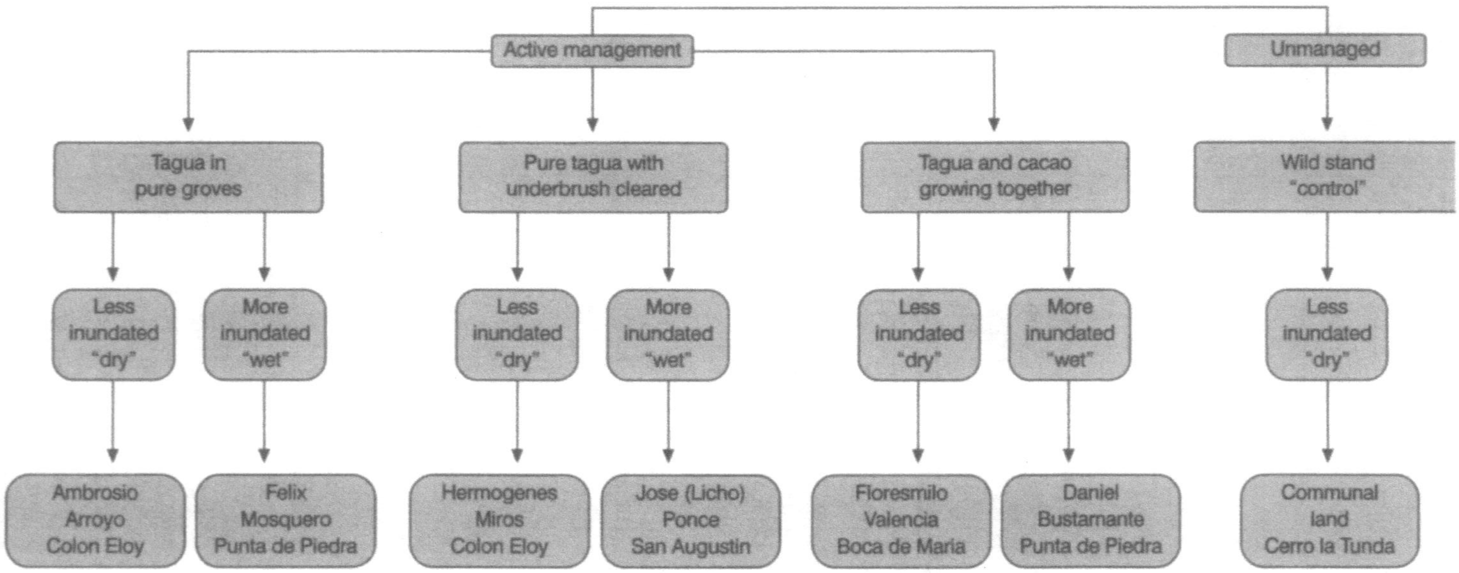

Figure 5.9 Local tagua management methods in Comuna Rio Santiago-Cayapas, Ecuador. Each management type is practiced on both dry and wet sites. No wild tagua stands could be found on wet sites, since all such land is currently under management or no longer holds tagua. The bottom row of the diagram lists the locations and landowner where demographic plots were established to sample each management method. (Reproduced by kind permission from Velasquez Runk, 1995.)

worth of demographic data clarifying the ecological sustainability of local tagua palm management. The parabiologists gained familiarity with the research methods and collected data accurately and efficiently. Production of a tagua management manual and other educational materials add to the project's effectiveness in promoting ecologically sound local management of tagua.

5.4.6 Demographics over time (level 3)

A single demographic census can give a clear picture of the health and vitality of a plant population, but this is still a snap-shot of the population at a single point in time. Where demographic trends are clear, such as when a light-demanding tree cannot regenerate under its own canopy, a census at a single point in time may be enough to discern the population's future barring management intervention. For more subtle demographic trends, however, following individuals over time and recensusing populations at intervals may be necessary.

Marking quadrats to ensure their relocation and recensus can be a challenge. In choosing the material you will use for the stakes or other markers, consider how much time will pass between censuses. Metal and plastic stakes are long-lasting, but understandably may be removed by people when these versatile materials are scarce and hard to obtain in nearby communities. For plots that are recensused every few months, stakes can be made from decay-resistant wood, bamboo, or the hard central shafts of palm fronds. Where sites have abundant stones or rocks, these can be piled into cairns or painted to denote plot locations. While not a substitute for on-the-ground marking, global positioning system (GPS) technology can facilitate plot relocation by helping survey teams to navigate back to the location of a previously established plot.

There are also computer programs available that model the demographics of plant populations over time. Management teams able to collaborate with university-based researchers may have access to such models, which can compensate for a lack of long-term field data. Computer modeling is also a powerful tool for combining other relevant information known about a species, such as that gained from life history investigations and vegetation assessment.

5.4.7 Life history investigations (level 3)

Even after monitoring populations and habitat dynamics, there may be additional traits of a plant that are important for its persistence but that remain hidden. Population monitoring, for instance, may fail to reveal a plant's dependence upon a specific set of disperser animals that are hunted locally and in increasingly short supply. As a result, it is important to know as much as possible about the complete life history of plant species that are conservation priorities. Life history investigations are not so much an activity of monitoring *per se* as they are a part of baseline information gathering. To assess where the knowledge gaps are for priority species, managers can pose the following sets of questions.

- Where does the plant grow? What environmental conditions and habitat associations does it favor?
- How does the plant reproduce? What are its pollinators? What are its dispersers? What are their life histories and current status?
- What is the natural lifespan of the plant? Is it subject to prominent diseases or pest attacks? How does it respond to such problems?

Specific life history traits of plants that can help to answer the above questions are summarized in Table 5.5, along with useful research methods. Some of these traits may have been addressed by preliminary investigations or studies by other researchers. In addition, local residents may be knowledgeable about the responses of certain plants and habitats to disturbance and other external impacts, such as

Table 5.5 Sample life history traits useful for monitoring plant resources (adapted from Menges, 1991)

Trait	Ways to assess
Reproductive variability	Measure annual/seasonal fruit and seed productivity per plant, per unit area
Longevity	Mark and follow individual plants over time
Survivorship	Establish plots to measure germination success, mortality rates per year, or per plant life stage or size class. Construct life tables
Microhabitat requirements	Measure small-scale variation in seasonal and daily temperatures, rainfall, soil moisture, soil composition and associated vegetation at the microsite level
Disturbance ecology	Historical analysis of fire, floods and other disturbance mechanisms at sites where plant occurs. Sample vegetation at sites with a known disturbance age and history
Reproductive strategies	Mark and monitor individual plants to document pollinators, timing of flowers and fruit production, and seed dispersal
Mutualisms	Observe animal behavior and its effects on plant reproductive success, germination, establishment, growth and persistence; also note associations between target species and co-occurring plants and fungi
Inbreeding tolerance	Pollination trials (hand-pollinating an individual plant and excluding subsequent pollinators) can be used to assess self-compatibility.

changes in human activities, weather pattern fluctuations, or invasions of exotic species [175]. This knowledge may encompass quite subtle phenomena, such as pollination or dispersal mutualisms between plants and animals.

Traits not previously assessed or any life history questions that cannot be clarified by local residents should be the focus of follow-up investigations. In one example that illustrates how understanding life history traits can help to guide conservation, ecologists analyzed the relationship between populations of African fig trees and the wasp species that pollinate them. Figs (genus *Ficus*) have one of the most unusual pollinator relationships of all plants. Due to the specialized morphology of fig inflorescences, each fig species is utterly dependent upon a single tiny species of wasp for pollination. The wasp lives out virtually its entire life cycle in the fig inflorescences and cannot survive for more than a few days away from a fig tree that is not in flower. The complexity of the fig–wasp mutualism is compounded by the fact that in most fig species, individual trees produce all of their flowers at once. In a given population, however, tree flowering is not synchronized; individual trees flower at different times throughout the year.

Ecologist Judith Bronstein and colleagues [33] censused the flowering patterns of fig trees in a population of *Ficus natalensis* along the Invindo river in Gabon over a four-year period. The researchers measured wasp pollination rates and then employed a computer program to model the population dynamics of figs and wasps over time. Using this approach, they estimated that between 95 and 300 trees are required to maintain pollinating wasp populations for four years or more. A minimum of 800 acres would be needed to maintain such a population of *F. natalensis* at its natural density. This study is an example of a **population viability analysis** that calculates a species' **minimum area requirements**, discussed further in section 4.3.3.

5.5 Summary

Together, these monitoring and evaluation approaches offer different ways of gathering information and learning about plant resources and the habitats and natural communities they inhabit. In some cases, such as determining past successional patterns at a site, managers may look to local residents for guidance and consultation. Other situations may involve collaboration with botanists or ecologists who have the expertise to conduct studies that shed light on particularly puzzling aspects of a valued plant's life history. However monitoring and evaluation is carried out, the baseline information it provides is an essential part of the foundation for sound actions and decisions taken during *in situ* management. Monitoring and evaluation complete the circle of action begun when a management plan is formulated and point the way for a plan to adapt, evolve and remain useful and relevant for managers, conservationists and local residents alike.

6

Traditional agriculture and plant conservation

Figure 6.1 Woven floodwater weir, part of a traditional Tohono O'odham floodwater field, near Topawa village, Sonoran Desert, Arizona. The weir is placed across drainage channels near fields and used to capture nitrogen-rich detritus carried by water runoff from summer rains. (Photograph: G.P. Nabhan.)

At first glance, the idea that agriculture and wildlands conservation can be mutual allies may seem an incongruous concept. After all, there would seem to be no greater ecological contrast than that between an unruly mosaic of native forest or prairie patchwork and the geometric grainrows or monotone carpets of plantation trees that comprise modern agricultural landscapes. Yet tucked away in steep mountain valleys, at the end of rutted desert tracks, in small clearings amid jungle trees, a different agriculture persists, practiced by farmers with techniques refined from centuries of rigorous adaptation to local environments. In such settings, the distinction between nature and native agriculture often blurs and both systems become part of a single ecological landscape. In many cases, these landscapes support unique plant resources found nowhere else.

Our goal in this chapter is to demonstrate the potential for managing protected natural areas in concert with **traditional** agricultural systems (also referred to as **subsistence, small-scale, low-input,** or **peasant** agriculture) and the plant resources they contain. If nothing else, natural areas and traditional agriculture are geographic allies; both are most prevalent in rugged or isolated territory that is often considered marginal for mechanized monocultures, unattractive to land-extensive industry, and far removed from larger urban population centers. In many cases, the majority of residents in settlements bordering or surrounding protected areas are farmers, cultivating both fields and forests, and also using many plants and animals from surrounding wildlands.

The discussion that follows aims to provide protected area managers and conservation professionals with the background to recognize traditional agricultural practices, understand their conservation value and adopt *in situ* management approaches that can gain the support of agricultural communities. In particular, we focus upon answering the following questions:

- What factors differentiate traditional agricultural practices from modern agroindustrial practices?
- What phytogenetic resources contained in traditional agricultural landscapes are of particular interest to society, and to conservationists in particular?
- What factors contribute to the maintenance or loss of such resources from traditional agricultural systems?
- How can agricultural plant resources and their traditional management be documented and quantified?
- How can natural resource managers and conservation professionals assist traditional agricultural communities to address and solve problems they face?

6.1 Defining traditional agriculture

The label 'traditional' implies activities that have been practiced and refined over many generations among a community or culture. In using this term with reference to agriculture, we refer to managed food, forage, or fiber-producing lands exhibiting a collection of co-occurring traits, listed below [45, 49]. Keep in mind, however, that not every traditional farm, garden, orchard, or managed forest plot will exhibit all of these traits.

- **A dominance of human and animal labor rather than mechanized labor.** Where farmers grow crops in small plots on steep or uneven terrain, the most efficient cultivation method may involve no more technology than a simple digging stick or hand-pushed plowshare. Where plots are larger and plowable, typically oxen, buffalo or other draft animals are used. Machines such as tractors and threshers are rarely employed.
- **Cultivation of a wide range of crops and multiple varieties of each crop for varied purposes.** The many species and varieties grown on a traditional farm have a variety of uses and include not only fruits, grains and vegetables for food, but also medicinal plants, spices, cash crops, ornamentals, building materials and fuelwood. The diversity of crop varieties in traditional agricultural fields can be spectacular as well. Expert Andean farmers grow up to 50 potato varieties on a single farm, and within a single village the total can reach 100 varieties [3].
- **Intercropping prevails over monocropping.** When farmers grow many kinds of crops, they often **intercrop** or interplant them in a single field or plot (Figure 6.2). The swidden plots of certain indigenous cultivators in tropical moist forests can contain upwards of 100 species of plants growing together in the same field at the same time [5]. Intercropping (also referred to as **polyculture**) often lends a disorderly look to traditional agricultural plots. Yet this apparent unruliness masks farmers' systematic interspersing of crops, which often promotes crop traits that interact to benefit all intermingled plants (Box 6.1).
- **Less reliance on commercial pesticides, herbicides and fertilizers.** In many areas, farmers cannot easily obtain or afford commercial pesticides, herbicides and synthetic fertilizers. Many aspects of traditionally farmed plots, such as intercropping and cultivating crops in small disjunct patches, serve to minimize pest and weed infestations and thus reduce the need for chemical controls. Where farmers apply fertilizers, it is typically animal manure or composted plant material (so-called green manure) rather than commercial nitrogen–phosphate fertilizer. In swidden or slash-and-burn agriculture, soil fertility is maintained by allowing a cultivated plot to lie fallow and revert to natural vegetation. Fallow periods can last anywhere from three to 30 or more years, depending on the time needed to restore soil fertility and on land-use pressures. In the Western Ghats of India, farmers evaluate the soil fertility of fallowed lands by observing the grass species that recolonize the fallows. Two grass species are especially watched for: one

Figure 6.2 Intercropped agroforestry plot near Una, Bahia, Brazil, where black pepper (*Piper*) and cacao (*Theobroma*) are being cultivated under peach palm (*Bactris*) and Brazil nuts (*Bertholletia*). (Photograph: R.P. Guries.)

Box 6.1 The benefits of intercropping

Scientific research has verified the benefits of an ancient example of intercropping, the maize (*Zea mays*), bean (*Phaseolus vulgaris*) and squash (*Cucurbita* spp.) polyculture practiced by Mesoamerican farmers for millennia. Agroecologist Steve Gliessman [88, 89, 90] reports that in Tabasco, Mexico, maize yields in traditional polyculture are as much as 50% higher than in monoculture plots. This over-yielding trend holds true even with corresponding increases in monoculture planting densities [9]. Although beans and squash each produce lower yields in polyculture than in comparable monocultures, the summed yield for all three crops tops that of any of the crop yields in monoculture (Table 6.1).

Table 6.1 Yields of maize, beans and squash in monoculture vs. intercropping at Cardenas, Tabasco, Mexico (adapted from Gliessman, 1990)

Crop	Density (plants/ha)		Yield (kg/ha)	
	Monoculture (best)	Polyculture	Monoculture (best)	Polyculture
Maize	66,000	50,000	1,230	1,720
Bean	64,000	40,000	740	110
Squash	7,500	3,300	430	80
Total		93,330		1,910

Note that while the productivity of beans and squash drop when intercropped with maize, the total polyculture yield is about 55% higher than the best monoculture yield.

Gliessman's studies have revealed several possible explanations for this polyculture's increased productivity and yield stability. Beans intercropped with maize produce more root nodules per plant, suggesting an increased capacity to fix nitrogen, which can then be uptaken by the maize plants. Squash plants appear to benefit beans and maize through weed control: the wide, thick, horizontal squash leaves cast a dense shade over exposed soil, reducing weed establishment and growth. Squash leaves also produce allelopathic compounds that are leached out by rainfall and may inhibit weed growth. Interestingly, however, the reason most commonly offered by farmers for why they intercrop is not that of yield stability; their expressed objective is more often that of reducing dietary monotony [209].

indicates a moderate return of soil fertility and a second only appears when soil fertility has recovered fully to pre-cultivation levels [252].

- **Greater reliance upon rainfall or local water flows.** Traditional Tohono O'odham farm plots in the Sonoran Desert are watered using only surface runoff from summer cloudbursts [150]. This runoff is exceptionally rich in nitrogen, for it flows through watersheds dominated by legume cover. Agroindustrial farms in the region, by contrast, depend on water from deep subterranean aquifers and large-scale diversion from the region's few permanent rivers. The surface irrigation water is also less salty, since the soils in O'odham *ak-cin'* or runoff farming are lower in alkalinity and richer in macronutrients than other desert soils. Some of these soils rank among the most fertile known.
- **Greater use of locally adapted crops.** Before the advent of intensive agricultural breeding programs in the 20th century, virtually all crop varieties were developed and maintained *in situ* by farmers who, year after year, saved seed from their best-performing crops to replant or trade with neighbors. Even today, researchers estimate that at least 80% of all crops planted worldwide are from seeds that farmers have saved themselves [253]. Traditional farmers do obtain extra-local seed, but then select it to local conditions over many seasons. While **locally adapted** varieties (also known as **landraces**, *razas criollas*, or **folk** varieties) cannot usually match the sheer productivity of **high-yielding** or **improved** varieties under ideal conditions, they often display other qualities such as superior resistance to local diseases or insects; drought, cold, or heat tolerance; superior performance in low-nutrient soils; long storage shelf-life; and flavor – all highly regarded traits. As a result, farmers often retain at least some traditional landraces even after high-yielding varieties have been introduced into a region [39]. For instance, it is common throughout Africa to find farmers growing high-yielding varieties to sell in markets and traditional varieties for their own cooking pot. The end result is that modern and traditional varieties assume distinctly separate roles in the household economy [253]. The degree of market production depends primarily upon farmers' access and proximity to markets [39].
- **Management is primarily through principles and practices derived not from formal scientific inquiry, but from orally transmitted, communally held knowledge and observation.** Awareness of management principles and practices is typically passed on orally or by demonstration, from farmer to farmer within families and communities.

Finally, it is also important to clarify briefly what is **not** traditional agriculture. The extreme of non-traditional cultivation – also called modern, intensive or agroindustrial farming – is characterized by large field size, extensive mechanization, high inputs of chemical fertilizers and pesticides, and heavy dependence upon groundwater irrigation or large-scale river diversion when practiced in arid or semi-arid regions. Agroindustrial crops are typically grown in monocultures, utilizing either high-yielding improved varieties (many of which are **hybrids**), most prominently in the case of rice, wheat and maize; or non-native plantation crops such as rubber and oil palm in Southeast Asia, coffee and sugar in Latin America and cacao in West Africa.

6.2 Who practices traditional agriculture?

Experts estimate that at least 1.4 billion people worldwide – a majority of them in Asia – depend directly upon low-input, subsistence farming systems that display many traits of traditional agriculture [78]. Some of the most vibrant and persistent examples of traditional agriculture come from ethnic communities with a long history of land use and cultivation in a particular area or region. One well documented case is the farming tradition of Huastec Maya peoples, who live along the foothills of the Sierra Madre Oriental in northeast Mexico. Ethnobotanist Janis Alcorn [1] found that Huastec Maya farmers manage a diverse assemblage of cultivated and fallow farm plots, forest patches and intricate home gardens, which together often contain over 300 plant species.

In certain circumstances, immigrant peoples also demonstrate ecologically compatible agricultural practices and have amassed a sophisticated knowledge of local natural history over relatively few generations. For example, the *caboclo* and *ribereño* communities of Brazil and Peru are known for their complex agroforestry systems, despite being relative newcomers to the Amazon basin [104].

Farmers practicing traditional methods also tend disproportionately to be women rather than men, a trend which reflects women's enormous contribution to agriculture. Recent estimates suggest women produce over 50% of the world's food in fields, gardens and orchards, with their regional contributions ranging from 30% in Latin America to 80% in sub-Saharan Africa [78]. Most of the crops women grow are destined for home consumption – precisely the crops most likely to be diverse, locally adapted varieties that do not get counted in most official agricultural surveys. As a result, women hold a wealth of knowledge about the uses and management of crop varieties and often have primary responsibility for storing seeds between harvests and selecting seed for planting. Women are also likely to be knowledgeable about wild plants gathered for food, such as edible 'weedy' species that colonize the margins and borders of farmers' fields [78, 181].

6.3 Plant resources in traditional agricultural systems

One advantage of the biosphere reserve approach to protected areas management is that it explicitly offers an opportunity to support and promote traditional farming systems [97, 159]. But what stake do natural resource managers and conservation professionals have in traditional agriculture? There are at least four reasons why managers seeking to conserve plant resources should be interested in working with communities that practice traditional farming patterns.

(a) Genetic exchange

Genetic exchange (also known as **introgression**) between crops and their wild relatives in nearby habitats has resulted in naturally hybridized plants and other unique phytogenetic resources in traditional agricultural landscapes [246]. Introgression between wild and cultivated races (Figure 6.3) has been documented for most major crops in their regions of origin, including rice (*Oryza*), wheat (*Triticum*), corn (*Zea*), potatoes (*Solanum*), chile peppers (*Capsicum*), sorghum (*Sorghum*), squash (*Cucurbita*) and tomatoes (*Lycopersicon*) [7, 183, 23]. Most cases of introgression are **intervarietal** (between domesticated and wild varieties of the same species, as between teosinte and corn) or **interspecific** (between plants of different species in the same genus, as with wild and cultivated potatoes). **Intergeneric** (between plant species in different genera) gene flow is much more rare. Transferred genes typically are most persistent in domesticated crop varieties, for they are rapidly selected out of wild populations.

In some cases introgression is deliberately encouraged by farmers, who recognize the potential genetic benefit their crops can receive from wild relatives, such as increased yield and hardiness in the face of diseases, drought and other stresses. Hybrid weedy and cultivated individuals of the squash *Cucurbita argyrosperma* in the northern Mexican state of Sonora show novel combinations of traits economically useful in the *pipián* or squash seed industry [155]. Some of

Figure 6.3 Introgression between wild gourds and cultivated squash (*Cucurbita* spp.) in a farmer's field near Oñavas, Sonora, Mexico. Note the intermediate fruit size in the introgressive field-edge hybrids. (Photograph: G.P. Nabhan.)

these field-edge hybrids produce smaller gourds with more seeds per plant than most cultivated squashes; others are sweet-fleshed with intermediate seed sizes easily selected by mechanical seed threshers. Introgression between maize and its close annual wild relative teosinte (both *Zea mays*) is another case where gene flow between wild and domesticated plants is encouraged by traditional farmers (Box 6.2).

Box 6.2 Swapping genes between maize and teosinte

Maize introgression with teosinte has been recognized by crop scientists since its description by the plant explorer Carl Lumholtz over a century ago, and archeological records suggest that indigenous Mexican farmers have been aware of the process far longer, perhaps for more than 5000 years [165, 247]. Economic botanist Garrison Wilkes [245] has observed Mexican farmers allowing teosinte 'to remain within or near corn fields, so that when the wind pollinates corn some natural crosses occur, resulting in hybrid plants'. Wilkes estimates that, all told, teosinte-corn introgression has occurred in more than 25 of the 50-plus traditional maize varieties grown in Mexico.

In the highland valley of Nabogame in northern Mexico, intensive livestock grazing has eliminated teosinte plants except on the steepest canyon slopes and stream channels – and in the fenced-off cornfields of Tepehuan Indian families. Tepehuan farmers claim that teosinte invigorates their maize crops, and genetic studies on teosinte and maize from Nabogame suggest a small but regular flow of genes between the two [65]. The introgressive hybridization that occurs is either not reciprocal (that is, genes from maize do not appear in the teosinte populations, perhaps due to phenological limitations), or maize traits acquired by teosinte are naturallly selected against [64]. As a result, Tepehuan maize cultivation does not threaten remaining teosinte populations, while teosinte does its part by enhancing the genetic diversity of maize.

(b) Habitat diversity

Local residents' historic agricultural practices have in some cases helped to shape the habitat mosaics that are the object of conservationists' interests [11, 190]. Nearly half a century ago, economic botanist Edgar Anderson [10] described the ecological dynamics of **hybrid habitats** between closed forests and open fields: the fallow plots, hedgerows, windbreaks, road margins, dump heaps and irrigation-derived wetlands that result from human management of agricultural landscapes. He observed that species which do not normally coexist can interact with each other in peculiar ways within these hybrid habitats. Where such hybrid habitats persist in a stable human-managed landscape, species richness and even speciation itself may increase over time [182, 190].

Such hybrid habitats are prime sites to search for unique phytogenetic resources, such as wild crop relatives. In Mexico's Sierra de la Manantlan Biosphere Reserve, the wild perennial relative of corn, *Zea diploperennis*, is most abundant in young secondary fallows of abandoned swidden maize plots (see Box 5.7). This plant would be much less likely to persist in the landscape if human agricultural activities and attendant habitat disturbance patterns were not present in the region [23]. While cases such as *Zea diploperennis* do not lessen the importance of maintaining undisturbed wild habitats for other species, they do illustrate how long-standing agricultural practices can have conservation value and need to be factored into management plans for conserving plant resources in protected areas.

(c) Crop genetic resources

Traditional cultivators are direct custodians of valuable genetic diversity in their fields. The accumulated pool of locally adapted crop varieties that traditional farmers plant and select seed from each year is more genetically diverse than the more narrow mix of intensively bred varieties that dominates industrial agriculture [79]. By constantly mixing varietal lines – sometimes with wild crop relatives through introgression – farmers reduce the probability that pathogens or insects will rapidly evolve to overcome any genetic resistance possessed by crops against these biotic stresses. In many cases, traditional farmers actively promote diversity in their fields by encouraging an unusual individual plant that displays traits different from the others with which it was planted [25, 250].

High-yielding hybrid crops, on the other hand, remain productive and resistant to pests and diseases only through chemical inputs and regular infusions of selected genes derived from traditional cultivars and wild crop relatives, accomplished via modern breeding programs [213]. Without access to the wellspring of genetic variation flowing ultimately from traditional farmers' fields, many high-yielding intensive monocultures would falter in productivity within a few crop generations [79].

(d) Subsistence value

In addition to fostering valuable plant resources, traditional farming systems represent the very means of survival for many rural residents and neighbors of protected areas. Francis [80] has estimated that as much as 15% to 20% of the world's food supply continues to be produced through traditional agriculture. Communities constrained from farming on their traditional land base may suffer severe economic and social hardship if this source of food or income is undermined and other subsistence alternatives are not available.

6.4 Why is traditional agriculture threatened?

The forces that erode the integrity of agricultural landscapes and threaten the plant resources they contain have a variety of origins. One pervasive problem is the loss of an adequate land base for practicing traditional agriculture. In some cases, small-scale farmers lose their entire land base when they are displaced from their traditional areas of residence by immigrant colonists, threat of violence from large landowners, or other social disruption [67]. The lands that displaced farmers end up with tend to be in remote areas with low fertility soils, rugged terrain, or other physical characteristics that limit the productivity and sustainability of agriculture.

Even when cultivators successfully retain their land base, internal social and ecological dynamics can radically reshape agricultural landscapes. Population growth may spur increasingly intensive land use as family farming plots are subdivided among offspring, fallow periods are shortened, and steep slopes and other marginal areas are brought under cultivation. In some situations these trends have led to increased soil erosion, loss of soil nutrients, more frequent pest infestations and related problems. Such problems in turn decrease the productivity of fields and plots and ultimately reduce the area and quality of land available to farmers [232].

Yet while population growth contributes to many environmental problems, it does not automatically lead to the breakdown of farming systems. In the 1930s, the hills and plains of the Machakos region of southeast Kenya displayed extensive soil erosion, gully formation and rapidly degrading vegetation [73]. Home to the Akamba people – who at the time practiced a combination of extensive livestock herding and small-scale cultivation – the Machakos region appeared at the limits of its human carrying capacity. Yet as the population of Machakos has subsequently increased fivefold, environmental conditions have actually improved, with large reductions in soil erosion and improvements in tree cover. A key reason for these changes appears to have been the ability of Akamba residents to adopt and apply more intensive land management techniques, such as bench terracing, that conserved soil and tree cover and could also support more people [73]. Population growth at Machakos has not been without side effects – native scrub habitat has diminished in area as agriculture has intensified, and preferred trees for wood carving are over-harvested and almost gone from the region [99] – but overall, the Machakos agricultural landscape has improved remarkably. This example demonstrates the importance of approaching each region, village, or project area as a unique situation, with its own blend of historical, cultural and environmental variables [255].

6.5 Genetic erosion

As cultivation systems and rural societies change, wild crop relatives, landraces and distinctive crop gene pools can be lost all too easily from agricultural landscapes. We have observed that loss of genetic resources occurs in several distinct patterns. These are presented in Box 6.3 and illustrated with examples from the US/Mexico Borderlands.

Box 6.3 Patterns of crop genetic resource loss

Widespread replacement of folk varieties with hybrids of a different crop species

Such replacement has been most prevalent in fertile, productive farmlands, such as lowland river valleys, that are close to urban centers and other major markets [4]. For instance, in the Río Yaqui valley of northwest Mexico (the region's breadbasket) native corn, beans and squash have been replaced by extensive stands of wheat, tomatoes and safflower grown for export markets.

Widespread replacement of folk varieties with introduced hybrids of the same crop

This pattern of genetic resource loss has also been heaviest in fertile regions accessible to large-scale markets. The Pima Indians of the Sonoran Desert grew a landrace of cotton known as Sacaton Aboriginal until the early 1900s. It was then replaced by a variety called Long-staple Upland, originally smuggled into the United States from Egypt. This cotton proved so productive and its products so readily accessible among Pima farmers that it rapidly made the folk variety obsolete.

Genetic swamping of a folk variety by an introduced variety

When Mammoth Russian Sunflower became a common garden crop among the Hopi Indians of Arizona, it swamped out the indigenous Hopi sunflower variety. Even though the Hopi did not abandon or otherwise alter their sunflower crop, the introduced variety so contaminated the native strain that the genes which made the indigenous sunflowers useful as a dye were lost.

Erosion of crop gene pools due to reduced field sizes and smaller crop population sizes

Virtually no genetic variation remains among the food and fiber crop *Agave murpheyi*, once cultivated by a prehistoric Native American culture in the Sonoran Desert. All remaining populations are small, seed set is low, and vegetative (clonal) reproduction is the norm. Erosion of crop gene pools also occurs when farmers divide their land into commercial production (hybrid varieties) and subsistence production (traditional folk varieties). In the process, traditional variety gene pools shrink, and some varieties may be abandoned entirely and go locally extinct [39].

Disruption of gene flow between wild relatives and crops

This most often occurs when cropland areas are expanded, and field margins and other hybrid habitats favored by wild relatives are eliminated. This happened with commercial chile pepper production in Mexico, where wild chiles (*Capsicum annuum*) were once common in hedgerows along field edges.

A number of underlying factors that fuel genetic erosion in traditional farming systems [4] can be summarized as follows.

- Access to urban and export markets that favor modern varieties.
- Access to prime land (flat, well watered) suitable for modern varieties.
- Farmer demand for cash income.
- Access to credit and/or subsidies for seeds, fertilizers, pesticides, irrigation and other inputs needed to grow modern varieties.
- Shortages of labor needed to cultivate traditional crops.
- Problems in saving seeds of traditional varieties.
- Crop failures due to war, social unrest.
- Crop failures due to drought, floods, or other climatic events.
- Long-term climate change.

The common feature linking these factors is that they cause farmers to change what they plant and how they manage their lands in ways that lead to genetic resource loss. In some cases, farmers may abandon traditional crop varieties because of events beyond their control, such as drought, floods, or social unrest. For instance, when severe drought struck Zimbabwe during 1991 and 1992, many small-scale farmers could not reap adequate harvests and had no choice but to consume the seeds they normally saved for planting. Subsequently, seeds of many traditional varieties were scarce and difficult to obtain [233].

Involuntary abandonment of traditional crops can also be triggered by long-term changes in environmental conditions. In certain parts of Senegal the climate has become markedly drier over the past several decades, such that in many years the traditional rice varieties grown by women are no longer suited to local conditions. The result has been rapid adoption of introduced, earlier-maturing rice varieties that yield more reliably under the new growing conditions [161].

In other situations, shifts in crops and varieties cultivated may be largely voluntary responses by farmers to new opportunities they perceive for improving food security, increasing their cash income, or other benefits. Crop genetic resource specialists generally agree that the spread of a relatively few improved crop varieties this century – produced by formal plant breeding programs – has fueled the displacement and extinction of thousands of folk varieties worldwide. For instance [78]:

- As the proportion of wheat fields in China planted to folk varieties dropped from 81% in the 1950s to only 5% by the 1970s, the total number of traditional wheat varieties grown fell from nearly 10 000 to around 1000.
- In Korea, a survey of 14 home garden crops found that only 26% of the folk varieties grown in 1985 could still be found under cultivation in 1993 – an alarming figure, given that home gardens are usually rich sites for traditional crops.
- In Mexico, only 20% of the maize varieties documented in a 1940s survey are still grown today. Reasons cited for this decline include a decrease in the area of fields planted to maize and the replacement of maize by other cash crops.
- The United States began experiencing extensive genetic erosion earlier than most other nations. Between 81 and 95% of all apple, tomato, cabbage, pea and field maize varieties documented in crop registries before 1904 appear to have vanished entirely. They are neither grown commercially nor retained *ex situ* in gene banks.

For the most part, farmers with land suitable for high-yielding crops have adopted them voluntarily. A prime enticement of high-yielding crops is their income-generating capability, stemming from their greater yields and marketability. In addition, national and international development programs often generously subsidize the costs of the seeds, water, fertilizer, pesticides and herbicides necessary to grow hybrids [79].

Cases of genetic erosion also stem from logical responses by farmers to social or environmental constraints. For instance, farmers who face challenges in saving seeds of traditional varieties, such as pest or fungal contamination, may prefer to purchase seeds of alternative varieties if they become available in markets, rather than saving their own [81]. In other situations, local or regional labor shortages can lead to loss of crop varieties whose cultivation or processing are particularly labor-demanding (Box 6.4).

Box 6.4 Labor availability influences crop persistence

Geographer Karl Zimmerer [255, 256] has documented trends in the diversity of crop varieties cultivated in Paucartambo, a valley in the highlands of southern Peru where potatoes and maize are the traditional subsistence crops. During the 1980s, seasonal household labor shortages in Paucartambo increased substantially, due in part to out-migration of residents for wage employment during certain times of the year. These shortages contributed to farmers' abandonment of a fast-maturing 'precocious' potato species, *Solanum phureja*, whose poor storability and lack of tuber dormancy essentially demand continuous cultivation (Figure 6.4). The very rapid maturation time of precocious potatoes had long made them a valuable crop, but between 1967 and 1987 the number of households growing them dropped from over 50 to only six. Farmers who continued to cultivate precocious potatoes tended to be from wealthier households, who could afford to hire extra labor and fence their fields. Fencing had become necessary because farmers who had previously grown precocious potatoes began grazing cattle on their now-fallow fields, and the cattle's wanderings increased the incidence of crop damage.

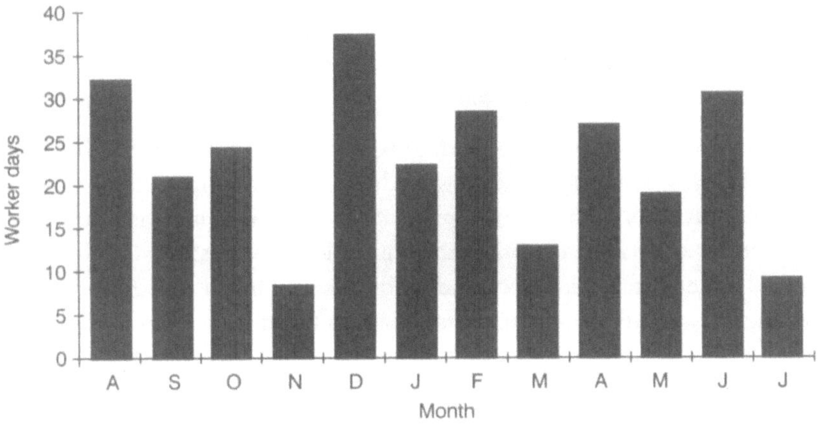

Figure 6.4 Monthly labor requirements (per hectare) for cultivation of precocious potatoes over a year. While more labor is required in some months than in others, the cultivation cycle is essentially continuous. The sample size for these data was five separate fields. (From Zimmerer, 1991, reprinted by kind permission of the American Geographic Society, Inc.)

As this brief discussion shows, the reasons why farmers alter their crops and cultivation practices are often complex and not always immediately evident [255]. However, agricultural researchers and conservationists are giving increased attention to methods that can help them to document and understand the diversity and loss of plant resources in agricultural landscapes. Such understanding lays the groundwork for *in situ* conservation, and the following sections look more closely at these methods and approaches.

6.6 Documenting agricultural plant resources

Without a doubt, many of the trends undermining traditional agriculture stretch beyond the direct purview of land managers and plant conservation professionals. Grassroots development projects, national agricultural policy reforms, family planning programs and similar political and economic initiatives are essential actions in the struggle to sustain diverse agricultural systems. Nevertheless, land managers and plant conservation professionals can make an important contribution by documenting the plant resources present in agricultural landscapes (particularly those in or adjacent to protected areas), drawing attention to trends and patterns that threaten these resources, and working directly with farming communities to develop workable solutions.

In discussing how to document and understand plant resources in agricultural systems, we focus on two points:

- Describing the patterns of fields, fallows, hedgerows, wetlands and forest plots present in traditional agricultural landscapes, and the ways in which these areas are managed and impacted by local residents. Such habitat mosaics often support wild crop relatives, ethnobotanical products and many other plants harvested by rural residents.
- Surveying and monitoring the traditional crop varieties cultivated within farmers' fields. The social and ecological dynamics critical for conserving landraces and diversity within crops are distinctive and unique, and come with their own set of challenges and opportunities for conservation. Of course in practice, agricultural landscapes and landraces are linked together in many ways, and conservation approaches for one may benefit the other as well.

6.7 Diversity and dynamics of agricultural landscapes

Gathering information on the kinds of habitats present in agricultural areas, their ecological functions and the useful plants they contain, is one way to identify priorities for conservation action in agricultural landscapes. First, let us briefly review several methods discussed in previous chapters that have application here. The rural appraisal methods outlined in Chapter 3 are tailor-made for obtaining an initial picture of the value that local residents place on different agricultural landscape elements. When carefully applied, semi-structured interviews, community mapping and field transects can provide information on how different habitat types are perceived and managed by farmers. Where more quantitative documentation of plant resources is desired, the inventory and monitoring methods outlined in Chapters 4 and 5 are useful. For instance, wetlands, corridors and patches of potential habitat for plants and both migratory and resident animals can be mapped with local residents to evaluate their extent, distribution and composition.

Rural appraisal methods such as ranking exercises also provide an entry to understanding the socioeconomic relationships that influence farmers' management of agricultural landscapes and the plant resources found there. Information on tenure and use patterns can be documented as well and incorporated into conservation planning. As resource managers, local residents and conservation professionals collaborate over time, it may be particularly useful to gather additional information specific to agricultural landscapes. In particular, we want to emphasize two areas:

- agroecological dynamics of farming systems;
- links between agricultural lands and resources and surrounding wild habitats.

6.7.1 *Agroecology*

Agroecologists use ecosystem concepts – originally developed by biologists for analyzing natural communities – to analyze cultivation practices and the agricultural environment. In particular, practitioners of agroecology seek to strengthen the productivity and health of agricultural systems by applying the ecological benefits gained from diversity and self-regulation, rather than applying external inputs like fertilizers and pesticides. Table 6.2 illustrates a number of components and elements that can be used to evaluate the ecological appropriateness of agricultural practices.

Table 6.2 Selected agroecological criteria useful for evaluating agricultural practices (adapted from UNDP, 1995)

Category	Criteria	Examples
Physical productivity and fertility	Soil erosion controls	Terracing Check dams Windbreaks Ground cover plantings
	Soil fertility maintenance/nutrient cycling	Crop residue composting Green manures N fixation Livestock wastes
	Maintenance of water supplies	Watershed cover Watershed area Aquifier recharge Irrigation efficiency
	Sustainability of inputs	External nutrient inflows
Agricultural and biological diversity	Genetic diversification	Locally adapted varieties/landraces Multiple varieties
	Community diversification	Intercropping Presence of: pollinators; natural pest enemies; detritivores and decomposers
	Landscape diversification: temporal	Crop rotations Fallow cycles
	Landscape diversification: spatial	Intercropping Agroforestry Hedgerows/field margins Patches of native vegetation

Social/cultural appropriateness	Sources of innovation	Locally initiated vs. external Application of indigenous knowledge and skills Continuity with traditional practices
	Relevance for local agricultural realities	Farmer participation in test plots/experiments Local evaluation of results Farmer-to-farmer dissemination of results

With this perspective, agroecological studies can illuminate particular benefits that plant resources provide to farmers and help to answer the question of why a farming community has chosen to maintain a particular mix of plant species and varieties in their fields, kitchen gardens, forest plots, orchards and field margins. For instance, agroecological research has demonstrated how the diverse polycultures common in traditional agroecosystems can help to limit pest populations within farmers' fields (Box 6.5).

Box 6.5 Pest control benefits of traditional polycultures

One benefit of intercropping consistently revealed by agroecological studies is a reduction in the densities of insects and other pests that damage crops. In parts of Nigeria, plots planted only with cassava (*Manihot esculentus*) are vulnerable to grasshoppers, and farmers will sow maize and sorghum with cassava until it is ready to be harvested [2]. In Tlaxcala, Mexico, fields where maize was intercropped among *sabino* (*Juniperus deppeana*) and *aelite* (*Alnus firmifolia*) trees had only one-sixth the density of pestiferous *Macrodactylus* beetles found in monocropped maize fields after 15 weeks [8]. In Southeast Asia, paddy rice cultivators traditionally maintained carp and other fish in the flooded paddies. By eating the eggs, larvae and adult insects that land on the water, fish can reduce pest levels by as much as 90%, while also fertilizing the paddies and providing farmers with a ready source of protein [128].

Agroecologists have at least two explanations for how polycultures control pests. First, when pests are specific to a single crop species, it may be harder for the pest to locate its target species and disperse among host plants. Second, a polyculture may support higher densities of pest predators than does a monoculture. When squash in Tabasco, Mexico, was grown in combination with maize and beans, herbivorous caterpillars that fed on the squash were more likely to host parasitoid wasps than those caterpillars feeding on squash in monoculture. Polycultures appear able to host more pest predators due both to increased variety of predator food sources (other insects, pollen and nectar) and increased availability of microenvironments and refugia [125].

Agroecological analysis also embraces a variety of social and economic approaches to understanding agricultural systems. Geographer Susanna Hecht [102] writes that agroecology, unlike traditional ecology, has arisen within a broad interdisciplinary perspective because:

> Agricultural strategies respond not only to environmental, biotic and cultivar constraints but also reflect human subsistence strategies and economic conditions. Factors like labor availability, access and conditions of credit, subsidies, perceived risk, price information, kinship obligations, family size and access to other forms of livelihood are often critical to understanding the logic of a farming system.

As a result, agroecological concepts can help to identify and quantify links between social factors and evident environmental trends, such as increased soil erosion or a diminishing fallow period, that are likely to influence the long-term stability of agricultural landscapes. Agroecological investigations may also suggest alternative practices by which farmers can retain important landscape patterns, such as hedgerows and secondary habitats for associated wild crop relatives, that would otherwise be lost. With its emphasis on farmer knowledge and the great variability and local specificity of farming systems, agroecological investigations are readily combined with participatory research methods and direct farmer involvement in conservation projects.

6.7.2 Links between cultivated lands and wildlands

One important focus for agroecological investigations is on the biological and physical interactions between different habitat types. Such interactions are what bind farmers' plots, secondary habitats and surrounding areas of native habitat into a single landscape [6] and we term them **reciprocal effects**. Through these links, events in a nearby wild habitat can influence processes in a farmer's field, and vice versa.

Understanding reciprocal effects is important for *in situ* conservation because it provides a full picture of the benefits and drawbacks that various habitat types present for farmers. Some reciprocal effects are readily evident, as when an elephant herd tramples villagers' fields while migrating between forest patches. Others are more subtle yet no less influential, such as when forest margins and hedgerows support more diverse pollinator populations, leading to increased fruit tree productivity in a farmer's orchard.

Specific reciprocal effects that may be particularly useful agroecological indicators for conservation professionals and managers are:

- soil erosion and/or deposition of sediment loads downstream from fields;
- streamflow surges (during storms) and ebbs (during droughts or periods of heavy irrigation);
- aquifer recharge or flow rates in artesian springs;

- water quality with regard to turbidity, nitrates, phosphates, pesticide residues, dissolved oxygen and other factors;
- 'nick points' of small but intensive land disturbance which aid in the establishment and spread of invasive plants;
- corridors of forage or nesting habitat for migratory species;
- microhabitats (shady orchards, artificial wetlands created by water leakage from irrigation canals) for plants and animals with special habitat preferences;
- pest and disease outbreaks;
- crop raids by wildlife;
- altered wildlife densities in nearby habitats due to local hunting pressures;
- altered densities of useful wild plants in nearby habitats.

The dimensions of these effects will vary in each particular case, depending, for instance, on the scale and intensity of agriculture and on the extent of the shared boundary or **edge** with wildlands. In addition to general system-level links, additional reciprocal effects between wildlands and cultivated lands may be worthy of documentation. Introgressive hybridization between crops and wild relatives is one of these effects (Box 6.6).

Box 6.6 Documenting and quantifying introgression

Introgression can be documented through several approaches. The first step is to talk directly with farmers, and then sample field crops and farm produce for apparently unusual intergrades between crops and wild relatives (see Figure 6.3). In addition, the potential for introgression in an agricultural landscape can be assessed by documenting which potentially interbreeding relatives of crops might be present in and around fields. Understanding the **phenology** of wild relatives and crops can also have predictive value, since both species or varieties must flower at the same time in order for cross-pollination to occur [203].

The case for introgression can be strengthened further by documenting actual **pollen transfer** between crops and nearby wild relatives [116]. Experiments can determine whether hybridization between two taxa is physically possible, for instance by deliberately moving pollen from the anthers of the flowers of one species, to the stigma of the flowers of another. The recipient flower is then subsequently monitored for fertilization, seed development and seed fertility. Most of this work has been done on insect-pollinated plants, but the techniques used are potentially transferable to wind-pollinated plants as well. Some researchers have used non-toxic dyes to stain pollen grains on a flower, and then observed other flowers for the presence of transferred stained pollen. Other studies have used fluorescent powders as pollen surrogates, dousing the reproductive parts of certain flowers in powder and then tracking its distribution to other flowers by pollinators [116]. Fluorescent powders are favored because they glow vividly under ultraviolet light, readily revealing their presence. Pollen transfer investigations are also useful for establishing pollen dispersal distances within a plant population and rates of pollen transfer. Studies have found that dye dispersal distances are more restricted than pollen, but even pollen exchange tends to occur primarily within a population. The immigration rate of pollen from one population to another is usually less than 1% [71].

There are potential limitations to be aware of when interpreting pollen transfer results. In some cases, pollinators have been found to avoid certain dye and powder colors, so it may be necessary to test several colors to find an appropriate one. Moreover, **pollen flow** is not the same as **gene flow** [72]. Even if pollen transfer has occurred, the pollen may not be viable. Furthermore, ecological factors may subsequently limit germination, establishment and consequent reproduction of hybrid offspring.

Finally, molecular techniques that directly measure gene transfer or gene flow between plants will give the most accurate picture of introgression. However, the equipment needed to test seed paternity directly is expensive and requires laboratory facilities. Collaborative work with universities or crop breeding centers is the best avenue for accomplishing genetic studies of introgression. Studies using genetic markers frequently show a greater amount of gene exchange than would be predicted by observations of pollinator movements and ranges [206]. If so, introgressive gene flow may occur between populations that appear at first to be isolated by distances between fields, elevational gradients, forest barriers and other landscape features.

The impact of agricultural activity on the local fauna is another important link to consider between agriculture and wildlands. Many protected areas with potential for being managed to support useful plant resources also have a mandate to protect and sustain wildlife populations. As a result, a key issue for many protected area managers to consider is how agricultural landscapes influence the local fauna and how negative impacts might be mitigated (Box 6.7).

Box 6.7 Wildlife in traditional agricultural landscapes

Traditional agricultural landscapes can provide valuable habitat for many animal species. Agroecosystems in arid regions, which essentially concentrate water to provide food resources for humans, often prove attractive to insectivores, foliage feeders and seed consumers [188]. In tropical forests, old fields and fallows tend to have high concentrations of favored wildlife food plants. Many tropical forest cultures practice a subsistence strategy known as **garden hunting**, where hunters specifically target and manage old swidden habitats to concentrate game populations [66, 127].

At the same time, traditional agriculture is not without impacts on native wildlife. Agroecosystems tend to be most representative of secondary or disturbed habitats, and are unlikely to support all animal or plant species that occur in unmodified natural habitats. For instance, traditional garden forests of Sumatra are exceptionally diverse tree crop plantings that mimic native forests in structure. However, when ornithologist Jean-Marc Thiollay [226] compared the avifauna in Sumatran garden forests with that of nearby primary forest, he found that bird species richness in all garden forest types was 12–62% lower than in primary forest. Garden forest characteristics potentially responsible for this reduced bird diversity include: fewer microhabitats due to lower average tree height, foliage volume and vertical structure; reduced plant species richness, implying a less diverse food supply; and high densities of nest predators.

In addition, wildlife hunting by farming communities can be substantial, and subsistence hunting for home consumption is widespread in many rural regions [189]. Hunting pressures are particularly strong where markets exist for wildlife meat or 'bush meat', as is the case in many central African countries [120].

Even with these problems, traditional agroecosystems tend to remain more compatible for native species than intensive agricultural landscapes. While Sumatran garden forests are less diverse than primary forest, they still support far more birds than do nearby monocultures of oil palm and other crops, which typically host only three or four bird species [226].

Similarly, in a study of irrigated Sonoran Desert farmland, ethnobiologist Amadeo Rea [187] found that traditional Pima Indian fields held much greater bird diversity than nearby mechanized fields, due to plot size, crop density, higher weed diversity and the absence of pesticide use.

Managers can work with wildlife biologists to quantify the abundance of wildlife and assess the impacts of agricultural practices and hunting (Appendix A). The most common type of field survey used for censusing wildlife is a strip transect, along which direct observations of animal species and indirect evidence (dung, tracks, vocalizations, feeding signs, territorial markings) are recorded. Interviews with local hunters and surveys of nearby markets can also reveal preferred game species, amount of hunting pressure, degree of hunting done by locals versus outsiders, and other variables. Together this information can serve as a basis for negotiating directly with communities on strategies for conserving the fauna found in and near traditional agricultural landscapes.

6.8 Landrace diversity and variability

Crop landraces have received much attention over the years from agronomists, plant breeders, geneticists and other academic researchers. Where their research has had implications for germplasm conservation, it has primarily addressed *ex situ* maintenance. As a result there now exists a substantial knowledge base that details how to collect, manage and conserve the genetic diversity of landraces in seed storage laboratories, genebanks and other *ex situ* facilities [95].

By comparison, the formal information base needed to conserve landraces *in situ* is in its infancy. The techniques to build this information base come from a variety of disciplines. The pages that follow demonstrate how conservation professionals can gather information on crop varieties to answer two questions fundamental to *in situ* landrace conservation:

- What is the diversity of crops cultivated in the area (village, buffer zone,watershed, etc.) being surveyed?
- Why have farmers retained or abandoned their locally adapted varieties?

6.8.1 Landrace diversity patterns

Landrace variation and diversity can be examined at radically different levels of detail. Thus the first step in assessment is determining the scale at which landrace sampling will occur. Studies have quantified the diversity of landraces cultivated by one family (see below), a community [40], a series of communities within a single region [40] and even across an entire country [129].

For surveys focused at the family or community level, or that are long-term and systematic, landrace samples and information about their cultivation are typically collected directly via farmer interviews and visits to individual farms. For instance, over a 20-year period, ethnobotanist James Nations has regularly inventoried the crop assemblage maintained by one Lacandon Indian farmer, Juan Camino Viejo, in Chiapas, Mexico. Although Juan's mix of crops varies from year to year, the snapshot-like surveys conducted by Nations cumulatively provide one of the best documentations of a single farmer's stewardship of diversity through time.

For regional studies where time and resources are limited, researchers often concentrate on surveying local marketplaces and recording the varieties available for sale. Such market-based surveys can then be supplemented by a random sample of farmer interviews to assess the proportion of varieties not showing up in markets. For both market and farmer-based surveys, care must be taken to collect information over the course of the full harvest period for any given crop – varieties can differ dramatically in their time of harvest. A snapshot survey at a single point in time is likely to capture only a subset of the landraces actually under cultivation.

A landrace survey for conservation purposes typically follows an ethnobotanical approach (as was detailed in Chapter 4) and involves collecting data on the names and numbers of crop varieties under cultivation. Relevant morphological characteristics should be recorded for each variety, along with other qualities that local residents may use to distinguish between varieties:

- palatability;
- aroma;
- cooking time;
- resistance to pests in field and in storage;
- time to maturation;
- storage longevity;
- ceremonial value;
- other features of cultural significance.

When studies measure landrace diversity in terms of morphological characteristics and number of varieties perceived – and maintained – as distinct entities by farmers, they are measuring the physical (or **phenotypic**) expression of genetic diversity rather than actual **gene frequency** levels. As mentioned in Chapter 4, there are a number of laboratory methods available that can more precisely measure the actual genetic variability of landraces. The most widely used laboratory method for assessing genetic diversity in crops and crop relatives has been protein electrophoresis, although systems using actual DNA measurements are seeing increasingly widespread application. Such direct genetic measurement techniques require laboratory facilities and tend to be relatively expensive. But where conservationists and resource managers have the opportunity to collaborate with university researchers and government agricultural research programs, laboratory-based genetic assessments can be useful for conservation planning (Box 6.8).

Box 6.8 Genetic surveys and Andean potato diversity

A research team led by Stephen Brush [40] has studied the diversity of potato varieties (*Solanum* spp.) cultivated by highland farmers near Cusco, Peru. The Andes mountains are the center of diversity for potatoes and are the site of their original domestication approximately 6000 years ago. The highly variable topography, rainfall, soils, daily temperature fluctuations and other environmental conditions encountered by farmers in the Cusco region have fostered a diverse potato gene pool, with the average farmer growing 10–12 different landraces, representing up to seven different potato species.

To understand the local distribution of genetic diversity in potato landraces, Brush and his colleagues randomly sampled 610 tubers in eight fields near the community of Paucartambo. Using electrophoretic techniques, they examined the diversity of **alleles** (different copies of a gene) and allelic combinations, or **genotypes**, exhibited by each potato sample (equal to one tuber) across four gene locations, or **loci** (Tables 6.3, 6.4). This provided a very detailed picture of the genetic landscape within local potato varieties.

Table 6.3 Number of alleles and population heterozygosity index of four genes in field samples of potatoes at Paurcartambo, Peru (from Brush *et al.*, 1995, reproduced by kind permission of Blackwell Science, Inc.)

Gene locus name	Heterozygosity index	Number of alleles
Pgi-1	0.29	3
Mdh-1	0.63	4
Got-1	0.07	3
Got-2	0.25	3
Mean	0.31	

Their findings have interesting implications for the conservation of genetic diversity among potato varieties. For alleles, there tended to be high levels of diversity within fields but low diversity between fields (i.e. the same alleles were present from field to field in varying numbers). This pattern suggests that germplasm collection efforts do not have to be extensive in order to capture most of the genetic variation in potato varieties – even merely collecting on a single farm might suffice for this task.

However, a very different pattern is evident for genotypes. There is a substantial number of very rare genotypes, each found in only one or two fields. This suggests that collection and *ex situ* conservation of all potato genotypes present in farmers' fields in Paucartambo would require an extensive effort. An *in situ* conservation approach may be more feasible in this case.

As mentioned earlier, population genetic inventories on this level are likely to be undertaken only for species of demonstrably great economic or societal value. Nevertheless, this study of potato gene diversity demonstrates the kind of questions and issues that can be resolved using a population genetics approach.

Table 6.4 Frequency of genotype combinations in field samples of potatoes at Paucartambo, Peru (from Brush *et al.*, 1995, reproduced by kind permission of Blackwell Science, Inc.)

Field ID Number	Very rare (< 2%)		Rare (2–5%)		Common (5–10%)		Abundant (> 10%)	
	No. of genotypes	Percentage of sample	No. of genotypes	Percentage of sample	No. of genotypes	Percentage of sample	No. of genotypes	Percentage of sample
1	5	3.6%	1	3.8%	–	–	3	93.0%
2	3	4.0%	2	6.7%	1	8.8%	3	81.3%
3	8	8.5%	4	13.8%	1	6.9%	3	70.8%
4	–	–	4	8.9%	–	–	1	91.1%
5	–	–	2	5.6%	3	16.7%	1	77.7%
6	–	–	7	16.7%	2	16.7%	3	66.7%
7	–	–	–	–	–	–	3	100.0%
8	–	–	–	–	–	–	2	100.0%
Average per field	2	2.0%	2.5	6.9%	0.9	6.1%	2.4	85.1%
Total of all fields combined	24	14.6%	3	8.0%	0	0	3	77.4%

6.8.2 *Landrace cultural contexts: understanding farmers' decisions*

As we have already seen, there are many reasons why farmers choose to maintain or abandon particular crop varieties. Identifying the particular factors at work in the farms, communities or region of concern may require gathering information on cultural, economic, environmental and agroecological subjects. Direct interviews and conversations with farmers are essential at this stage and can yield a wealth of detailed information. For instance, when ethnobotanists Jose Andrade Aguilar and Efrain Hernandez Xolocotzi [12] conducted a study relating common bean (*Phaseolus vulgaris*) landrace diversity patterns to conditions of agricultural production in the Mexican state of Aguascalientes, they used farmer interviews to gather information on a range of environmental and socioeconomic variables. First they recorded local names of bean varieties and collected samples of each. In addition, they interviewed farmers regarding the following variables:

- ecological conditions for bean production (rainfed vs. irrigated);
- number of years of continuous use of a plot;
- cultivation method (monoculture, intercropped with maize, other);
- size of cultivated plot;
- sowing date;
- type of power used for cultivation (human, animal and/or machine);
- use of fertilizers, insecticides or uniform improved cultivars;
- grain yield.

Andrade and Hernandez also examined regional climate records and geological maps to establish the mean annual temperature, mean annual precipitation, frost frequency, soil type and general topography for the sites they surveyed in Aguascalientes. By combining all of this information, they were able to describe clearly the local trends that influenced farmers' individual decisions on which bean varieties to maintain and which to discard.

Information-gathering can be standardized by employing the rural appraisal methods presented in Chapter 3. These methods are ideal for untangling the complex mesh of socioeconomic and agroecological factors that influence farmer selection, cultivation and maintenance of landrace varieties. Research and conservation teams from organizations such as the International Potato Center (CIP) are now employing rural appraisal techniques in surveys of landrace diversity and associated indigenous agricultural knowledge (Box 6.9).

Finally, useful information for interpreting landrace diversity patterns can also be gleaned from secondary information sources such as social surveys, government-sponsored environmental profiles, meteorological records and other regional studies. It is of course best to review these secondary materials prior to conducting fieldwork, so that they provide a context for information gathered directly from farmers.

Box 6.9 Rural appraisal applied to landrace surveys

Sweet potatoes (*Ipomoea batatas*) are a staple crop of highland communities in the Indonesian province of Irian Jaya, which covers the western half of New Guinea. Although sweet potatoes were originally brought to the island from the Americas, New Guinea farmers have developed a tremendous diversity of folk varieties. In 1992 a research team from CIP, an affiliated organization called UPWARD, and Cenderawasih University initated one of the first varietal surveys of Irian Jaya sweet potatoes that both sampled local landraces and used rural appraisal methods to document farmer knowledge related to sweet potato cultivation [177].

The survey focused on the highland farming communities of Iray and Sururey. From reviewing background material on sweet potato cultivation in Asia, the survey leaders knew that women have prominent roles in managing the crop, and they made sure to include a woman researcher on the team. In addition, they recruited two professionals with social science backgrounds to accompany the four biological scientists on the team.

The unit of study was the farmer's plot, and the team collected samples of all sweet potato varieties present in each farmer's plot that was surveyed. To document local knowledge concerning each variety and the ecological, cultural and economic context of farming, the research team drew up a **topic guide** (Appendix C) that summarized the data they hoped to obtain from local assistants. Once in the field, the researchers revised and added to the topic guide as important new information became available.

To collect information for the topic guide, the researchers conducted semi-structured interviews with individual farmers and with key assistants such as village elders and local political leaders. The former provided data specific to crop varieties, while the latter furnished information on village history, land use trends, seasonal work patterns and related subjects. To further document such information, interviewees completed transect diagrams, village resource maps, seasonal labor calendars, personal biographies and other tools.

The team also held group interviews. These took place in the field immediately after collecting samples of sweet potatoes growing in a plot, as well as in more leisurely evening gatherings organized by village elders. In both cases, group interviews emphasized comparison of sweet potato varieties, to elicit farmer perceptions of different varietal qualities.

The research team met each evening to review the topic guide and compile the information gained from the day's interviews and discussions. Each compilation was correlated with the sweet potato samples collected that day, using a common reference number. After this fieldwork, the researchers reflected on their experience and arrived at several conclusions that illustrate the challenges of rural appraisal work [177]:

- The greatest amount of information on cultivars came when several cultivars were compared and contrasted together, rather than when each was discussed in isolation.
- Gathering data on cultivation methods and farmer opinions of landraces was more time-consuming than the actual collecting of sweet potato samples and documentation of their basic **passport** information (species and variety name, date and location collected, etc.).
- Women farmers were a most valuable source of information, but tended to defer to men and let male villagers speak for them in mixed company, which was difficult to avoid in the fields. Interviews of women farmers were quite successful when the female researcher conducted them in womens' homes, without men present.
- Group interviews in the field yielded a wealth of information and were dynamic and informal, yet these aspects also confounded attempts to record farmer opinions and evaluations. Evening group interviews featured more focused discussion that made for easier recording, but such meetings also tended to be dominated by prominent village members. This led to a bias in the information gathered towards the views of male elders.

6.9 Conserving plant resources in traditional agricultural landscapes

While federal natural resource agencies and international agricultural research institutions bear the formal mandate for conserving agricultural plant resources *in situ*, their task again comes down to working with local residents who inhabit and manage agricultural landscapes. Indeed, the case for local collaboration is all the stronger for agricultural plant resources because they are so closely dependent on human management for their continued persistence. As we review possible approaches to *in situ* conservation in agricultural settings, we will emphasize their links with community development and local management issues, including those already covered in Chapter 3.

6.9.1 Conserving agricultural habitats

Working with local residents to conserve the habitat of useful plants in traditional agricultural landscapes is likely to involve a combination of approaches. Several approaches can be illustrated by examining a conservation proposal for corn's wild annual relative, teosinte (*Zea mays*), in its range in Guatemala [244].

In Guatemala, teosinte occurs in two distinct areas. The first, in the country's northwest corner, appears to be related to Mexican teosintes, but a second population area in southeast Guatemala is morphologically and genetically unique and may represent an ancestral line for the maize–teosinte species complex. Garrison Wilkes originally surveyed these populations in the mid-1960s and, in collaboration with staff from the Guatemalan Institute for Science and Technology (ICTA), he revisited teosinte stands in 1991. The team found dramatic population declines during this time period, approximately 75% in southeast Guatemala and 90% in the northwest. Once an abundant element of rural field margins and fallow plots, teosinte in Guatemala now occurs only in scattered, disjunct stands.

Several factors appear to be hastening the decline of teosinte. Increased grazing pressure from cattle is a major impact, as livestock relish the sweet, green leaves and stems of teosinte that appear after summer rains. Many farmers have begun to grow coffee and other crops that are not compatible with teosinte. Intensified land use has also led to reductions in the amount of fallows, hedgerows and other secondary habitats favored by teosinte. In northwest Guatemala, according to Wilkes [244], 'teosinte has almost disappeared except at the margins of distribution where corn is still the major crop'.

The *in situ* conservation program advanced by Wilkes and ICTA for teosinte has several habitat-related proposals. The first is lending support for traditional land use practices and related cultural customs in areas where teosinte is still prevalent. The link between cultural traditions and management of habitats is important to emphasize. Sacred groves, ceremonial grounds, sites of religious shrines and other culturally significant locations often protect patches of uncultivated or less disturbed habitat within agricultural landscapes. Managers may have opportunities to work with community religious leaders, cultural awareness committees and other local activist groups to help to ensure that cultural traditions important for maintaining habitats continue to be passed on between older and younger generations of community members.

Wilkes and ICTA also recognize the need to bolster local awareness of the uniqueness, importance and increasingly threatened status of teosinte. In general, this can be done through community education projects, particularly those implemented creatively and appropriately (see section 3.3.7). For such education efforts, it may be possible to collaborate with individuals, such as village elders, who are particularly knowledgeable about the plants of concern and well respected locally. Through story-telling, community meetings and other events led by both local experts and visiting conservation professionals, the importance and status of imperiled plant resources can be disseminated. In this way, prominent plant resources like teosinte give value to the marginal habitats they occupy and strengthen awareness for habitat protection.

A third component is the need to reduce the **intensity** of use and improve agricultural productivity on land where teosinte populations are in trouble. This point is a major one and we explore it further in the following section.

6.9.2 *The role of agricultural sustainability*

Sometimes natural resource managers may seek to work with farming communities on reducing land use pressures on fallow lands, communally held forest patches and other secondary or semi-wild habitats in agricultural landscapes. Sustainable agricultural initiatives can identify innovations to improve productivity on areas that are already intensively modified or degraded. Such an approach complements community education efforts by giving farmers alternatives to past management patterns that have fueled land degradation and loss of habitat diversity.

One way to monitor the ecological sustainability of agricultural practices is via participatory research techniques which resource managers and farmers may employ in concert. For instance, sites to measure soil erosion can be easily accessible to all participants and need not involve complicated procedures or equipment. One method of measuring soil deposition and erosion requires little more than several sturdy wooden or metal stakes. The stakes are placed at various distances apart along the downstream or downwind side of a field and pounded into the soil so that 10 cm is exposed above ground level. On a seasonal basis, managers and farmers can read the degree to which soil has accumulated around the stakes, or to which erosion has further exposed them.

An agroecological approach also offers a good starting point for developing agricultural innovations that move towards sustainability and optimal use of cultivated land. Between 1981 and 1989, the rural development organization World Neighbors conducted a project to increase agricultural productivity in several villages in and around the community of Guinope, Honduras [77]. Working directly with farmers and extension agents, the project introduced improvements in cropping practices, including novel crop rotations, intercropping with leguminous forage plants, organic fertilizer applications and micro-irrigation systems. The project also trained farmers in soil conservation practices, emphasizing soil retention through grass barriers, rock walls and drainage ditches along the contours of slopes.

The impact of these practices on farm productivity was evident almost immediately. Crop yields – predominantly maize – rose dramatically among the 1200 farming families participating in the project, from around 600 kg/ha to over 1200 kg/ha in the initial years and over 2,000 kg./ha by the project's end [43]. As a result of the increased yields, farmers did not have to devote as much land to staple crop production as before. Some of this land has been left fallow and is reverting back to native pine forest, while other plots have been put into pasture and fruit trees [229]. In addition, pressure on remaining forest areas has decreased, since many farmers who previously cut trees for income during the dry season are now occupied with irrigated plots and soil preparation activities [77].

From an agroecological perspective, these improvements in agricultural productivity have led to a more diverse, less degraded agricultural landscape – one that may have greater potential to retain populations of useful plants than before. By reducing land use pressures, productivity improvements may also give more breathing space for programs to increase local awareness of the value of plant resources in agricultural landscapes.

While agricultural innovations should have a solid agroecological grounding, they must also meet the social, cultural and economic requirements of farming communities. Rural development experts Roland Bunch and Gabino Lopez [43] have put these requirements another way, stressing that any agroecological innovation, to be sustainable, must either increase crop yields or decrease farmers' costs, while at the same time not increase the risks involved. The best filter for screening potentially useful new technologies or management innovations is to seek farmer input and participation at every step of the development process. Community meetings, participatory action research, community exchanges and other tools for working with local residents can be particularly useful at this stage.

6.10 Conserving landraces: farmer-based approaches

Anyone who remains skeptical about the ability of small-scale farmers to serve as landrace conservators would do well to visit the province of Cotabato on the Philippine island of Mindanao. The low-lying, well irrigated rice paddies of Cotabato in the backyard of the International Rice Research Institute are an ideal showcase for the Green Revolution. By the mid-1980s they were widely planted to genetically uniform high-yielding rice varieties (HYVs).

These fields are very productive, but occasional episodes of heavy flooding can devastate vulnerable paddies. Following one flood in 1985, a farmer named Eulogio Sase, Jr from the town of Santa Catarina noticed that a few rice plants had survived the high waters in a field that he had planted with a high-yielding variety, IR-36. He was not even sure that the plants were that variety – they could have come from rice seeds transported in by the floodwaters – but when the plants matured, Mr Sase collected their seeds and replanted them several months later in a small plot. He carefully observed the plants' performance and selected seeds for several generations. Eventually he stabilized a population of this new rice variety, which he named 'Bordagol', meaning 'short, solid and strong' [24, 201].

This variety has turned out to have an appealing combination of traits. Although it yields 20% less than currently available high-yielding varieties, it requires half the chemical fertilizer inputs and is less suceptible to pest infestations. Furthermore, because the plant tillers profusely it is able to outgrow weeds, thus greatly reducing the need for weeding or herbicides. But perhaps the most valuable trait of Bordagol is its appealing palatability. It has a taste distinct from all other Philippine rice varieties and commands a superior price in regional markets. Not surprisingly, the popularity of Bordagol did not stop with Mr Sase. He distributed the seed to other farmers and Bordagol is now widely planted throughout Cotabato and other Mindanao provinces. Already some farmers are growing distinct selections of Bordagol specific to their own fields [24].

This account has several guiding messages for *in situ* conservation. First, it shows that even today, it is farmers' keen interest in, observation of and selection for varietal differences that keeps crop assemblages diverse. Furthermore, it demonstrates that farmers have no hesitation about conserving crop diversity when it produces direct benefits for them. This perspective suggests that *in situ* landrace conservation can be advanced through collaborative programs that build on the knowledge and experience that farmers have towards landraces, and that it can increase the value placed on landraces within local farming communities.

It should be stated at the outset that *in situ* conservation approaches will be unlikely to conserve every last element of landrace diversity present in an agricultural landscape at any one time. For instance, as Bordagol spread among Mindanao farmers, it could have displaced traditional rice landraces – though the likelihood of this is small given that the region's paddies are dominated by high-yielding varieties. The more important point is that because crop selection, cultivation and replanting is a dynamic process, no *in situ* conservation approach can preserve every particular crop lineage that is documented in snapshot surveys. For this reason, we will focus here on conserving the **processes** that maintain diversity in farmers' fields. When these processes are in place and thriving, then farming systems will tend to sustain diversity within the suite of crops cultivated, regardless of the fate of particular varieties. At the center of these processes, of course, are the farmers themselves.

At present, the development of farmer-based approaches to *in situ* conservation is a grassroots movement led by non-governmental organizations, along with certain innovative national and international government agencies. Because their efforts are strongly influenced by each particular local context, they do not all follow a set pattern. However, their collective experiences suggest a number of cultural and ecological emphases (sections 6.10.1 to 6.10.3) for conserving traditional crop varieties *in situ*.

6.10.1 Keeping locally adapted crops valuable to farmers and farming communities

Projects that emphasize the unique qualities of local landraces can bolster their value to farmers and provide incentives for their cultivation. Even in areas where there has been extensive adoption of high-yielding varieties, some farmers usually maintain small stocks of traditional varieties [39]. For many farmers, their traditional seeds have unique value that high-yielding varieties simply cannot match. In addition, there is increasing documentation of cases where farmers are seeking to return to traditional varieties as a way out of the high-input treadmill

[25, 252]. The following points outline several ways in which *in situ* conservation initiatives can add value to landraces.

(a) Cultural value

In some cases, landraces have unique social or cultural values that provide incentive for their conservation. The Mende people of Sierra Leone include the distinctive red-hulled seeds of native African rice (*Oryza glaberrima*) in certain religious rituals. Introduced Asian rice varieties (*Oryza sativa*) are now widely cultivated in Sierra Leone for their superior yields and greater marketability, but the Mende do not consider them acceptable for traditional rituals. As a result, Mende farmers have a powerful cultural incentive to maintain stocks of native African rice [230].

To help to ensure that landraces retain cultural value, conservation professionals can work alongside community religious leaders, development groups and others with an interest in strengthening a community's or culture's traditional practices. In such situations, landrace conservation becomes inseparable from maintenance of cultural traditions.

Among Native American peoples of the southwestern United States, communities that have held on to traditional crop varieties have done so, in large degree, because of the important role the crops play in celebrating and reaffirming distinct cultural practices. The Hopi of northern Arizona continue to grow their own unique varieties of beans and corn that are essential elements of their communities' religious and social ceremonies. Modern varieties are unacceptable for ceremonial use and not just because of any particular foreign stigma attached. Experiments have demonstrated that Hopi gray lima beans and blue corn are exquisitely adapted for rapid germination in the sandy, nutrient-poor soil typical of Hopi farmlands. Modern hybrid corn or standard lima beans do not germinate with the same facility in Hopi soils and thus cannot fulfil their ceremonial roles as harbingers of fertility and successful harvest [146].

For many Hopi (and other Native Americans) traditional landraces are not maintained for reasons of physical subsistence or economic benefit. Rather, the old seeds are defining elements of culture and identity. As long as people – young as well as old – value and pass on the language, stories, myths, ceremonies and symbols of their culture, the seeds will not be lost.

(b) Social prestige

Other efforts have tried to give value to landraces by increasing the social prestige or status they bring to farmers. Some grassroots organizations have sponsored community and region-wide contests for the most diverse farm or greatest number of varieties cultivated of a single crop. Natural locations for such events include seed fairs and field days, which bring farmers together throughout a region to learn new agricultural techniques, share ideas and compare experiences (Box 6.10).

Box 6.10 Crop diversity contests

On certain days of the year throughout the Peruvian Andes, residents of communities scattered across different ecological zones congregate at established locations to celebrate religious festivals and sell goods after the harvest period. These regional fairs are prime opportunities to exchange information and ideas as well as goods. Certain fairs are known as events where farmers can exchange and obtain plants and seeds of the many indigenous crops and landraces. For instance, a fair in June in the Peruvian city of Cusco focuses on locally grown fruits, while another Cusco fair in December is dedicated to medicinal plants.

A group of Peruvian NGOs and local researchers realized that these regional fairs were ideal for promoting grassroots *in situ* conservation of traditional Andean crops [221]. They decided to organize contests at regional and community seed fairs to recognize and reward farmers who grow the most diverse range of crops and crop varieties.

The first contests were organized in regions known to be local centers of diversity for traditional Andean crops. Several months in advance of the fairs, contest organizers informed leaders of each community in the region of the chosen date and place, and explained the contest objectives and rules. At the fair, contest participants were registered and interviewed to document their crop species and varieties, the distinctive qualities of each variety (color, size, culinary properties, hardiness, etc.), and related agroecological information. Each entrant's harvest was then judged by a team of two agricultural technicians and two local farmers known for their expertise with crop diversity.

Awards were given both for varietal diversity within a single crop, and for diversity within general categories of crops such as 'tubers'. Winning farmers received prizes of tools and seeds, and official recognition at the fair, and certificates of participation were also distributed. According to Peruvian researchers Mario Tapia and Alcides Rosas [221], over 30 diversity competitions had been held at seed fairs by 1993, lending increased recognition and community status to farmers who continue to grow traditional crop varieties.

(c) Economic value

A major factor fueling a farmer's shift to high-yielding varieties is their marketability and ability to generate cash income for the farmer. Markets for traditional varieties are typically much more limited than markets for modern varieties. Commercial food companies tend to purchase only a narrow suite of modern varieties, because they prefer to process, package and sell a uniform, standard product. Certain people – particularly urban-oriented and younger consumers – may prefer modern varieties because they attach a stigma of backwardness or rusticity to traditional varieties [119].

Yet despite these drawbacks, traditional varieties have distinctive advantages that can, when placed in the proper circumstances, increase their appeal to all consumers – rural, urban and migrant populations. Consider the issue of nutrition. A growing body of evidence suggests that a diet based on diverse traditional crops can supply higher levels of many trace nutrients than a diet based on a few modern improved grain varieties [210]. As the cultivation of high-yielding grain varieties has spread in developing countries, researchers have documented rising iron, zinc and vitamin A deficiencies in people's diets. For instance, much of the high-yielding wheat grown throughout north Africa and the Middle East is notably deficient in zinc [210]. As consumers in a society become more concerned about nutritional problems, the appeal of traditional varieties may rise. In the southwestern United States, Native Seeds/SEARCH has worked with Native American communities to demonstrate how traditional foods can help to control diabetes and other diet-related illnesses now plaguing many native peoples (Box 6.11). In addition, efforts to develop greater cultural awareness and pride within ethnic groups may offer an opportunity to strengthen the appeal of landraces as links to one's past, one's traditions and (for migrants) one's homeland. Community and development projects may be able to build upon these qualities to increase consumer demand and markets for traditional crop varieties [119].

Box 6.11 Diabetes prevention by reviving the consumption of native foods

In the 1980s, one of us (GN) was told by an O'odham elder that our research on her tribe's traditional crop varieties was interesting, but not very useful since few people in her community farmed those crops any more. Instead, she asked, 'Why not use your research smarts to help us with this diabetes problem that is devastating us?' Today, about half of the Arizona O'odham over 35 years of age suffer from adult-onset diabetes; half a century ago, fewer than 4% of this population suffered from this nutrition-related disease. The cost of health care per diabetes sufferer has been estimated to be over US$8000 per year, and more than 250 000 individuals of indigenous descent in the US/Mexico border region are expected to suffer from this disease by the year 2000 [145].

The dramatic rise in O'odham diabetes frequency has taken place too quickly to be linked to the evolution of greater genetic predisposition to the disease. This suggests that something in the traditional lifestyle of the O'odham formerly prevented the expression of this genetically latent trait. While decreasing exercise levels have undoubtedly played a role, diet change appears to be a central factor, particularly by producing a greater tendency towards obesity; young O'odham men today weigh over 12 kg more than their counterparts of the same height measured around 1950. Our research has revealed that it is not merely that men today consume more calories than their counterparts half a century ago – it is the kinds of food which they consume that has changed so radically.

Guided by elders' reminiscences of plant foods that have fallen out of the O'odham daily dietary regime, we collected samples of historically important food plants under permit in protected areas and indigenous reserves. With the help of nutritionists Jannette Brand and Charles Weber [30], we determined that many of these native staple foods have high contents of soluble fiber, insulin and tannin. As a result, native foods more slowly release their sugars for digestion and absorption into the human bloodstream than do highly refined and processed foodstuffs derived from modern crops. When placed together in a diet with the same number of calories as found in a diet of refined highly processed foods, the native foods provided better control of blood glucose, cholesterol and insulin levels – all key factors in diabetes. Native Americans volunteering to reintroduce selected native foods back into their diets experienced rapid declines in blood sugar and cholesterol levels and began to lose weight. When placed on a regime of high soluble fiber foods and regular exercise, their vulnerability to diabetes was even further reduced.

As researchers, we not only published our findings in journals read by health professionals working in the O'odham region of Arizona; we also presented the same results in bilingual handout sheets, videos, slide shows and curriculum guides. We trained over 1000 health professionals and educators working in the vicinity of biosphere reserves, communicating the value of native food plants and encouraging land managers to allow the traditional harvest of renewable native plant parts (seeds, flower buds, stems) that local people could use in their treatment of diabetes. Currently, indigenous educators are doing most of the community outreach on how to incorporate these foods into a diabetic diet, and they do most of this teaching in their native language [145].

6.10.2 Increasing farmers' access to landrace diversity

Many cases have been documented where farmers are interested and eager to cultivate traditional landraces, but have lost their seed supplies. As previously mentioned, landraces are sometimes lost due to crop failure following drought, floods, or social unrest. In other cases, farmers become disenchanted with improved cultivars and seek a return to growing traditional landraces, but can no longer locate seeds. In the Philippines, high-input rice farmers in one community, facing stagnating yields, declining rice prices and rising chemical and fertilizer costs, decided to turn to organic rice farming. In making this switch the farmers had to look to remote villages for seeds, because rice landraces suitable for low-input cultivation had vanished entirely from their own community years before [25]. Below are several approaches that *in situ* conservation projects can take to increase the availability of landraces and farmers' access to them.

(a) Community seedbanks

One approach taken by grassroots organizations working on landrace conservation has been to help villagers to establish local seedbanks or nurseries for collecting, storing and propagating landraces. Some community seedbanks focus specifically on indigenous landraces, while others also incorporate suitable varieties that originated in other regions and commercially introduced varieties now saved locally. The essential components of a community seedbank usually consist of: a physical facility, such as a community building or nursery plot on common land, where seeds or seedlings are stored in bulk; and a volunteer committee or network of farmer conservators who perform essential seedbank tasks such as arranging for seed collection from farmers, overseeing seed storage, growing out collected landraces for multiplication and to ensure viability, and organizing seed distribution. One example of a community seedbank in action is presented in Box 6.12.

Box 6.12 Establishing a community seedbank

In the Indian state of Maharashtra, the Indian Academy of Development Sciences (ADS) has worked with farmers in four local districts to establish a community seedbank that can serve local farmers' demands for traditional varieties of rice [117]. ADS initiated this work when organization staff, in conversations and meetings with farmers, heard many farmers expressing their dissatisfaction with growing high-input rice varieties, due to stagnating yields, mounting pest infestations and the rising costs of inputs. These farmers were very interested in returning to traditional rice landraces, but had few sources of seeds for the old varieties.

ADS organized seed collection tours to locate, visit and obtain seed stock from the scattered farmers who still grew traditional rice varieties. In the first year these collection tours yielded 20 landrace varieties, which were sown and multiplied. After three years of seed collection, the ADS seedbank began to serve local farmers. Farmers are invited to visit the seedbank field plots as the rice grows and to place requests for seeds of desired varieties. The seed stock is then multiplied and distributed to farmers, in 1–2 kg amounts, via seed distribution camps attended by farmers. Farmers who receive seed are expected to deliver back to ADS 1.5–2 times the quantity of seed received, or else pass the same quantity along to another farmer.

The community rice seedbank has grown steadily since its inception and 'distributed seeds of nearly 60 indigenous varieties to more than 1000 farmers' during the summer of 1995, according to ADS Secretary Rajeev Khedkar [117]. The commitment of ADS to root the rice seedbank amid local farming communities is further evident by the makeup of project staff. Khedkar stresses that although agricultural professionals provide orientation and assistance, 'all activities – from seed collection, documentation, and maintenance of genebank, to characterization of cultivars – are being carried out by farmers'.

In addition to meeting farmers' seed needs, the conservation benefits of this kind of community seedbank are readily evident. The ADS seedbank now holds about 250 rice varieties, which is more than farmers are actively interested in growing at present – that number remains about 60 varieties [252]. However, by safeguarding varieties not presently in active demand for cultivation, the seedbank helps to make sure that those varieties will be available should growing conditions or farmer preferences change in the future. This may represent a point for collaboration between local and national seedbanks, since national seedbanks are geared more towards long-term storage. Regional *ex situ* institutions may want to concentrate on safeguarding these less demanded lineages, while letting community seedbanks take the lead on maintaining seed stocks of crop varieties currently in high demand by farmers.

(b) Traditional seed exchange networks

Most agrarian societies have local institutions and networks that facilitate exchange of seeds between farmers, increase farmers' access to landrace diversity and help them to cope with loss of seeds during calamities. Many such networks operate informally within and between communities. Farmers may exchange seeds while visiting relatives, travelling to markets and conducting other social interactions. Other seed exchange traditions are more structured and codified (Box 6.13).

Box 6.13 Traditional institutions for seed exchange

Many traditional seed exchange institutions operate on a local or village level. For instance, in some rural villages in Zimbabwe all residents are expected to contribute seeds annually to the community seed stock [233]. The community stock is redistributed to all village members at the start of the planting season, following a blessing ceremony conducted by the local chief. This traditional seedbank serves to ensure that no community member goes without seeds for planting. Moreover, it gives all villagers access to the full range of seed diversity present in the immediate vicinity.

Other traditional seed exchange institutions operate on a more regional scale. During the 1980s much of Ethiopia, including the region of Tigray, suffered both an extended civil war and widespread drought and famine. Tigrean farmers were completely isolated from national or international agricultural assistance, and had to depend entirely on local resources. Harvests were very meager, and community leaders attributed this in part to the generally poor quality of seed available to farmers. However, nearly every village and town held at least one farmer recognized as being an excellent seed selector, who usually had seeds superior to the general pool of planting material.

Starting in 1988, Tigrean community leaders came together and enlisted local expert farmers to help communities in establishing local 'banks' for seed exchange [25]. These banks were managed by assemblies of elected community members. The banks worked by offering superior-quality seeds – as selected by local experts – to needy farmers in the community, with the understanding that any farmers so supplied would return part of their harvest to replenish the bank. The first banks proved so successful that they were soon replicated by nearby communities. By the end of the war in 1991, local seed exchange banks had spread across a stretch of Tigray containing over 4 million people [25].

The banks have proven extraordinarily popular, and continue to function as Tigray rebuilds a state government and national society. Though not established for *in situ* conservation purposes *per se*, the banks have great potential as a foundation for community seedbanks to safeguard landrace diversity, participatory breeding programs to improve landrace quality, and other components of a landrace conservation program.

A useful first step for conservation professionals working in a region is to clarify the degree of establishment and current status of seed exchange networks. Interviews, historical timeline exercises and community meetings can be used to assess whether traditional exchange networks are thriving or dwindling, and whether farmers view lack of access to traditional seeds as a major constraint on their agricultural systems. Subsequent conservation efforts, such as promoting community seedbanks and sponsoring regional agricultural fairs and markets, can then be combined with traditional exchange institutions to increase their effectiveness and presence.

Regional grassroots organizations oriented towards conserving crop diversity are in a good position to link these issues together. An Australian-based seed conservancy, the Seed Savers Network, has been working to strengthen traditional supply networks for seeds and other planting materials throughout the Pacific region [208]. The Network is collaborating with local counterparts in Tonga, Western Samoa and the Solomon Islands to share information on seed selection, production, cleaning and storage methods, to establish locally based networks for locating, producing and distributing high-quality planting material and to promote community education activities such as demonstration gardens. Native Seeds/SEARCH, which focuses on traditional crops of the southwestern United States and northern Mexico, has similar programs underway with Native American communities in both countries. Further information about both organizations is listed in Appendix A.

6.10.3 Tailoring formal breeding programs to maintain landrace diversity

Formal plant breeding is a major international activity and we have discussed how improved crop varieties produced through formal breeding have displaced many highly diverse landrace assemblages. Yet there is no intrinsic reason why formal breeding approaches cannot support *in situ* conservation. What is required is that formal breeding programs recognize the value of locally adapted crop varieties tailored to the on-farm conditions faced by farmers. This adaptability springs foremost from a crop's genetic diversity. It is the driving force behind the diversity seen in traditional landraces. The easiest way to implement such a breeding program is to have extensive farmer input into the plant breeding process. When farmer participation has been a major part of breeding programs, the result has been improved crop varieties (**cultivars**) that retain high levels of diversity.

Formal breeding is essentially just an extension of the selection which farmers have done for millennia by saving the best seeds of each harvest for replanting. Plant breeding intensifies this process by controlling how individuals (or crop lines) mate with each other, and by arranging specific mating patterns, or **crosses**. The most basic form of plant breeding is **mass selection**, wherein varieties selected from one year's harvest are purposefully crossed with each other, rather than sown and allowed to recombine at random in a field [191]. Farmers and breeders alike practice mass selection widely.

Farmers who move beyond simple mass selection to pursue more elaborate breeding procedures on their own are the exception, but do occur. In any farming community there are likely to be individuals regarded as expert farmers by their neighbors, relatives and other associates. Such farmers may grow a wider diversity of crops than other households. They may be individuals to whom neighboring farmers turn for hard-to-find seeds or sage advice on planting and managing their crops. In some cases, local expert farmers practice their own plant breeding trials. For example, a Nigerian farmer from the state of Oyo on his own initiative collected around 20 varieties of cassava grown in the region and planted them on his land. He then observed and compared their growth and yield, and also undertook field trials using cassava plants grown from seeds – cassava is normally propagated clonally, using stem cuttings [81]. All of these approaches are widely used by plant breeders. Studies have also reported grassroots plant breeders at work with sorghum [230], rice [201] and maize [245].

During this century, professional plant breeders have developed a dazzling array of techniques to advance plant breeding beyond mass selection. Yet only in the past few years have breeding programs begun to incorporate a fundamental component: farmer perspectives on what they desire in a crop variety. Crop improvement programs that incorporate extensive farmer input fall into two primary categories: participatory varietal selection and participatory plant breeding [241, 242]. Each approach has important ramifications for *in situ* landrace conservation.

(a) Participatory varietal selection (PVS)

In participatory varietal selection, farmers conduct trials and select crop varieties for planting from a range of cultivars that have already been refined by scientists in formal breeding trials. Participatory varietal selection begins with identifying the traits that farmers in a particular community or region require and value most in their crops [241]. Rural appraisal methods are useful at this stage and can be complemented by having farmers pre-select varieties from trials of potential candidates grown at a research station or an on-farm study site.

Once farmers' requirements for crops are clarified, then a search can be made for cultivars, such as those just emerging from breeding programs or previously released with limited distribution, that appear to meet the desired traits. This pool of candidate cultivars is then subjected to comparative trials and selection in farmers' fields. The degree of involvement that farmers have at this evaluation stage can vary according to the project design. Possible approaches are presented in Table 6.5. When correctly implemented, PVS trials should identify those improved cultivars that will produce well for farmers under their local conditions and constraints.

(b) Participatory plant breeding (PPB)

Participatory plant breeding incorporates farmer involvement earlier in the breeding process, in some cases starting with initial crosses of landraces and other diverse varietal lines. As with PVS, all evaluation and selection at each stage is done by farmers in on-farm trials; the difference is that in participatory plant breeding, farmers are working with landraces and not just with cultivars that have already gone through an intensive formal breeding process for particular traits. Participatory plant breeding approaches include collaboration between farmers and breeders using mass selection to enhance landraces, crossing cultivars identified through PVS with landraces and even crossing landraces, cultivars and high-yielding varieties.

Table 6.5 Approaches to participatory varietal selection (PVS) (from Whitcombe and Joshi, 1996b, reproduced by kind permission of International Plant Genetic Resources Institute)

Method of farmer participation	Means of evaluation	Typical institution involved
On-farm trials, researcher-managed; replicated design	Yield data	Research
On-farm trials, farmer-managed with scientist's supervision; several entries per farmer; replicated design	Yield data	Research
Farmer-managed trials, replication across farmers, one variety per farmer	Yield data; farmer's perceptions	Research; extension; NGO
Farmer-managed trials, replication across farmers, one variety per farmer	Farmers' perceptions	NGO; extension; research
Farmer-managed trials; no formal design	Informal, anecdotal	NGO with limited resources; extension

(c) PVS vs. PPB: advantages and drawbacks for conservation

Although participatory varietal selection and participatory plant breeding are relatively similar in application, there are important distinctions between the two approaches [242]. PVS tends to be easier to set up and implement than PPB and seed supplies are usually more readily available for PVS. However, participatory varietal selection may in some cases exacerbate local genetic erosion, such as when a single improved cultivar that meets farmer approval displaces several traditional landraces. If PVS-identified varieties are displacing older improved cultivars, then overall genetic erosion is likely to be much less severe. Indeed, participatory varietal selection can lead to genetic enhancement when multiple cultivars are adopted by farmers who before grew only a single variety (Figure 6.5).

Participatory plant breeding, while logistically more complicated than participatory varietal selection, is not likely to reduce on-farm genetic diversity significantly, even over the short term [242]. This is because it incorporates and draws directly upon the broad genetic base of farmer landraces and exposes farmers to a large pool of varietal diversity in the breeding process. In essence, PPB parallels the selection and experimentation process which farmers have always conducted on their crops. Of course, any breeding program carries the potential for losing unique genes as landraces are mixed and selected for over time, and PPB is no exception. As J.R. Whitcombe and K.D. Joshi [242] point out, participatory plant breeding will not automatically preserve all traditional landraces – that is a job for *ex situ* seed banking – but it will ensure that the crop varieties in farmers' fields are genetically diverse, locally adapted and full of evolutionary potential.

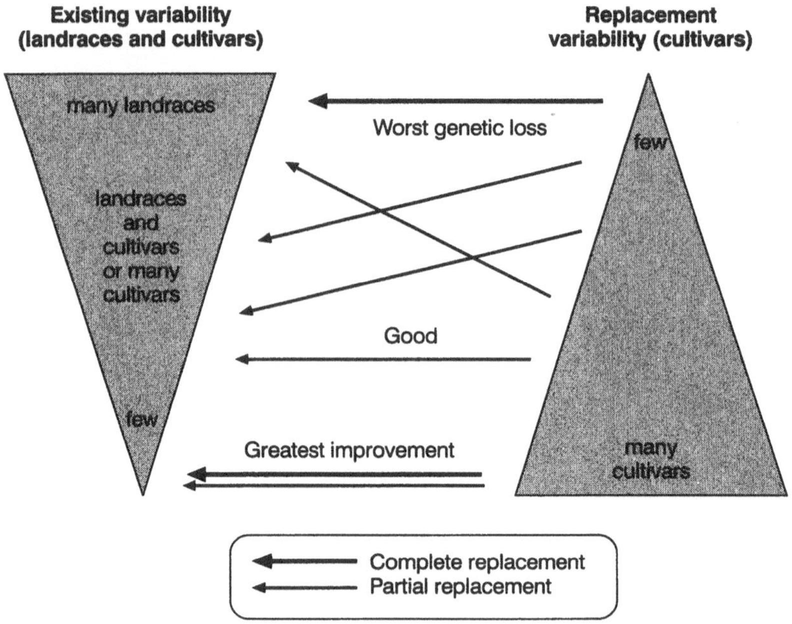

Figure 6.5 Potential patterns of landrace replacement by new cultivars from plant breeding programs. (From Whitcombe and Joshi, 1996a, reprinted by kind permission of the International Development Research Centre.)

6.11 Intellectual property rights: who benefits from seeds and other plant resources?

As local seed exchange institutions demonstrate, farmers and cultivators worldwide have traditionally shared crop varieties on a free or reciprocal basis [204]. Farmers have also shared their varieties widely and freely with economic botanists, agricultural resource specialists and other individuals who collect folk varieties for *ex situ* storage. Some of these samples have been used by plant breeders to develop commercial seed lines for which either the breeders or their commercial sponsors have proprietary use rights. The debate over how the economic benefits from such use should be distributed is one of several issues related to plant conservation that raise concerns about **intellectual property rights to resources**, or as they are more broadly known, **ways to value traditional ecological knowlege** [37].

Of particular concern are cases where native plants – particularly folk varieties and medicinal plants – and traditional knowledge about them are collected for one reason (such as *ex situ* conservation or biological inventory) but then used for an unrelated purpose (for-profit development of hybrid seed lines or pharmaceutical compounds). Unfortunately, many examples exist where enormous royalties have been accumulated by patent-holders on hybrid variety crop lines and pharmaceuticals, and by institutions sponsoring plant collection and evaluation, without any benefits returning to the local communities whose knowledge guided the initial investigation of the plant [79]. Although examples are arising where botanists, institutions and corporations are returning benefits to communities, many local people remain wary of sharing their knowledge of plants with outsiders, for fear of seeing resources usurped or others profiting at their expense.

The debate over intellectual property rights to landraces and other ethnobiological resources is actually between four camps [37]. First, there are **common heritage** proponents, who feel that no single cultured community should be compensated as if they are exclusive innovators of a plant resource, since historic multicultural efforts were responsible.

Instead, common heritage advocates feel that compensation should consist of funds held in trust for all local people, indigenous communities and rural residents, to help them with resource conservation and cultural preservation efforts. A second group uses the common heritage arguments to decline compensation payments to all parties except individual innovators who can demonstrate that they have trade secrets (such as a herbalist who knows how to use a particular combination of medicinal plants), or have done unique work to improve a plant resource (such as an expert farmer who develops a distinctive crop variety).

By contrast, a third group believes that **farmers' and foragers' rights** to royalties should be given to the communities from which both the plant resource and pertinent knowledge about it are derived. From this perspective, such rights should be legally equivalent to the **breeders' rights** granted to individuals or companies who formally patent a pharmaceutical compound or hybrid variety line. A fourth group believes that while the idea of farmers' and foragers' rights is fine in principle, actual evaluation of the relative contribution of each past and present community or group to traditional knowledge and resource stewardship is fraught with complications [37]. For instance, what happens when a valuable medicinal plant occurs across a wide range and is used by many different people in a variety of regions? Whose rights get counted and whose get excluded in determining compensation? Proponents of this fourth perspective feel that non-monetary benefits (such as assistance in health care, oral history preservation, or land protection) should be offered to every community prior to ethnobiological explorations or plant collections, regardless of whether or not any product commercialization is anticipated. Then, if a project is commercialized in the future, additional negotiations should be held to make the community a co-steward and benefit-sharer.

Debates between these four perspectives continue to produce further ethical guidelines applicable to all researchers, managers and conservationists involved in documenting and maintaining plant resources. In the meantime, the 1992

Biodiversity Convention sets clear mandates for the exploration of traditional crop varieties and other economic plants [78]. In general, it is important that:

- local residents and their representatives understand what their participation entails and what potential uses of their knowledge and products may result from their participation;
- local residents have a right to choose to be involved, to decline if they wish, to negotiate conditions for their participation (including confidentiality or review by the community of all or part of subsequent information transfers) and to renegotiate the terms of their participation should conditions change;
- local residents receive copies of any research results, in a language understandable to them, along with notices of and participation in any commercial developments emerging from their plants, products, or knowledge.

These and other tenets have been adopted by numerous non-government groups and certain international agencies working with the ethnobotanical resources of indigenous communities and local residents. Among academic professional organizations the International Society of Ethnobiology, the Society of Economic Botany and the International Society of Chemical Ecology have all adopted codes of conduct for scientists conducting ethnobiological research [59]. The United Nations Food and Agriculture Organization has adopted a Code of Conduct for Plant Collectors and their collaborators, and the Pew Scholars Program in Conservation has assembled a bibliography and clearing house on documents that discuss intellectual property rights, biodiversity and indigenous peoples (Appendix A).

6.12 Summary

Just as there is no clear ecological boundary between traditional agricultural lands and wildlands, *in situ* conservation in biosphere reserves and similar protected areas must encompass wild plants, field-margin species and locally adapted crops alike. Because management in agricultural landscapes depends upon collaboration with local residents at every stage, this work also blurs the distinctions between nature conservation and community development. Scientific approaches such as landrace diversity surveys or agroecological studies are invaluable ways of analyzing agricultural systems. But to stay relevant for management they must be applied to questions and issues that matter to the individuals who work the land, harvest the crops and otherwise influence the persistence of plants.

The need to provide alternatives to the modern social, economic and environmental trends eroding agricultural diversity will also link natural resource managers and conservationists with community development experts, grassroots plant breeders and others interested in issues of agricultural sustainability. All can offer valuable input and direction for managing plant resources and working to keep the diversity of agricultural landscapes intact, in place, and in use.

Appendix A
Further reading

This appendix presents further information on selected references and information sources that we found particularly useful in writing this manual. Our aim here is to highlight practical, comprehensive materials that will be most valuable for resource managers, conservationists and others working in the field on the daily tasks of conserving plant resources. Besides these specific materials, the publishing houses mentioned below are key source points for additional books on themes relevant for *in situ* plant conservation. We recommend contacting them directly for publications lists and information on their latest offerings.

Chapter 1

For information on the value of wild and traditionally cultivated plants, the work of Robert and Christine Prescott-Allen is among the most comprehensive. We particularly recommend *Genes from the Wild: Using Wild Genetic Resources for Food and Raw Materials* (1988), available through Earthscan Publications Ltd, 120 Pentonville Road, London N1 9JN, UK. An additional useful resource is a bibliography on wild foods entitled *The Hidden Harvest*, available through the International Institute for Environment and Development (IIED), 3 Endsleigh Street, London WC1H 0DD, UK.

The volume referenced in the References section as Falk, Millar and Olwell (1996) discusses the many challenges of *ex situ* collecting, propagation and reintroduction of rare plants. It is available through Island Press, 1718 Connecticut Ave. NW, Suite 300, Washington DC 20009, USA.

Chapter 2

One journal dedicated to illuminating the value of local natural resource management practices is the *Indigenous Knowledge and Development Monitor*, available from The Center for International Research and Advisory Networks (CIRAN/Nuffic), PO Box 29777, 2502 LT, The Hague, The Netherlands. In addition, the book *Biodiversity: Conservation, Culture and Ecodevelopment* (1991), edited by Margery Oldfield and Janis Alcorn, presents an illuminating review of many of the dimensions of traditional resource use and its application to conservation. It is published through Westview Press, 5500 Central Avenue, Boulder CO 80301-2877, USA.

Historical and sociological examination of the relationship between conservationists and local residents can be found in two papers referenced under Colchester (1994) and Pimbert and Pretty (1995). Both are published through the United Nations Research Institute for Social Development (UNRISD), Palais des Nations, 1211 Geneva 10, Switzerland. Two other excellent collections of essays and case studies along people-parks themes include *Natural Connections*, edited by David Western *et al.* (1994), available through Island Press (address above, under Chapter 1) and *Law of the Mother* (1993), edited by Elizabeth Kempf and available through Earthscan Publications (address above, under Chapter 1). An additional analysis of integrated conservation and development projects referenced as Wells and Brandon (1992) is available from The World Bank, 1818 H St NW, Washington DC 20043, USA.

Chapter 3

One of the best introductions to rural appraisal techniques can be gained through an informal serial called *RRA/PRA Notes*, published by IIED (address above, under Chapter 1). The rural appraisal workshop manual referenced as Freudenberger and Gueye (1990) is also available through IIED. An excellent presentation of rural appraisal techniques applied to participatory forest management, referenced as Poffenberger *et al.* (1992), is available through the Society for Promotion of Wastelands Development, Shriram Bharatiya Kala Kendra Building, 1 Copernicus Marg, New Delhi 110001, India.

The above works also cover participatory rural appraisal themes. In addition, an informative summary of the subject is *Implementing PRA: A Handbook to Facilitate Participatory Rural Appraisal* (1991), available through the Program for International Development, Clark University, Worcester MA 01610, USA.

A useful source on participatory action research and other community-based conservation approaches is referenced as Davis-Case (1990) and is published through the Forest, Trees and People Program, Community Forestry Unit, Forestry Department, Food and Agriculture Organization of the United Nations (FAO), Via delle Terme di Caracalla, 1-00100 Rome, Italy. The manual referenced as Bruce (1990) is also available through this program and contains in-depth information on tenure issues in natural resource management.

A very useful GIS program called CAMRIS is being made available to conservation organizations through an outreach program sponsored by the World Wildlife Fund. CAMRIS allows for full mapping capability, yet is designed to run on a standard personal computer and printer. For more information, contact George Powell, World Wildlife Fund, Apartado 83870, San Jose 1000, Costa Rica.

Chapter 4

Comprehensive reviews of ethnobotanical field methods include *Ethnobotany: A Methods Manual* by Gary J. Martin, available through the People and Plants Initiative or from Chapman & Hall, and *Selected Guidelines for Ethnobotanical Research: A Field Manual* by Miguel Alexiades, available through Scientific Publications Department, New York Botanical Garden, Bronx NY 10458-5126, USA.

Ecogeographic surveys are reviewed in greater detail in *Collecting Plant Diversity: Technical Guidelines* (1995), edited by L. Guarino and others, available through CAB International, Wallingford, Oxon OX10 8DE, UK. This book also contains a wealth of information on the collection and *ex situ* storage of plant genetic resources.

Many of the concepts behind participatory planning for natural resource management are covered in the manual referenced as Davis-Case (1990) (address above, under Chapter 3). A very thorough review of planning and monitoring for community-based conservation currently in the works is *Measures of Success: Designing, Managing, and Monitoring Conservation and Development Projects* by Richard Margoluis and Nick Salafsky – check on availability through the Biodiversity Support Program, c/o World Wildlife Fund, 1250 24th St NW, Washington DC 20037, USA, and through Island Press (address above, Chapter 1).

Chapter 5

Approaches to designing and implementing community-based monitoring are covered in a very accessible fashion in Margoluis and Salafsky, mentioned above under Chapter 4. The manual referenced as Davis-Case (1990), described above under Chapter 4, also covers some of the nuts-and-bolts of participatory monitoring and evaluation. The manual by Poffenberger *et al.* (1992) described under Chapter 3 has an excellent presentation of field monitoring methods that can be pursued in a participatory manner. Additionally, ecological aspects of monitoring forest management are covered thoroughly by Charles Peters in two manuals referenced as Peters (1994), available through the Biodiversity Support Program (address above, under Chapter 4), and – especially useful – Peters (1996), available through The World Bank (address above, under Chapter 2). For information on the tetrazolium test for seed viability, see the International Seed Testing Association (1985) International rules for seed testing. *Seed Science and Technology* **13**, 299–355.

Chapter 6

The best overview of agroecology is undoubtedly Miguel Altieri's *Agroecology*, republished in a new edition (1995) and available through Westview Press (address above, under Chapter 2). A book entitled *Agroecology: Creating the Synergisms for a Sustainable Agriculture*, published by the United Nations Development Program, also offers a broad introduction to the field and is available via UNDP, One United Nations Plaza, New York NY 10017, USA.

An engaging serial that covers agroecological topics – as well as grassroots agricultural themes in general – is the *ILEIA Newsletter*, available from the Information Centre for Low External-Input Agriculture, Kastanjelaan 5, PO Box 64, NL-3830 AB Leusden, The Netherlands. Issues related to sustainable agriculture in developing countries also appear frequently in a series of research papers sponsored by IIED, called the *Gatekeepers Series*. One of the best introductions to agroforestry and the role of trees in sustainable agriculture can be found in *Agroforestry in Dryland Africa*, by Dianne Rocheleau and others, available through the International Council for Research in Agroforestry, ICRAF House, PO Box 30677, Nairobi, Kenya.

A grassroots perspective on traditional agriculture and genetic resources is also elaborated in *Cultivating Knowledge: Genetic Diversity, Farmer Experimentation and Crop Research*, edited by W. de Boeuf and others and published by Intermediate Technology Publications, 103–105 Southampton Row, London WC1B 4HH, UK.

Pollination biology techniques that are applicable to studying patterns of introgression and other agroecological topics are reviewed thoroughly in Kearns and Inouye (1993), available through the University of Colorado Press, Niwot, CO. A discussion of techniques for monitoring wildlife can be found in the *Wildlife Field Research and Conservation Training Manual* (1993) by Alan Rabinowitz, available through the Wildlife Conservation Society, 185th St and Southern Blvd, Bronx NY 10460, USA.

Most of the technical information available on how to document and assess landrace diversity has been elaborated in reference to *ex situ* storage, such as in *Collecting Plant Diversity: Technical Guidelines* (address above, under Chapter 4). Additionally the journal *Economic Botany*, published by the Society for Economic Botany, PO Box 368, Lawrence KS 66044, USA, carries the findings of many researchers working to document and understand both cultivated and wild plant resources.

Two grassroots organizations with extensive experience in seed banking, networking and community-based conservation of rare crop varieties are Native Seeds/SEARCH and the Seed Savers' Network. You can contact Native Seeds at 2517 N Campbell Ave. #325, Tucson AZ 85719, USA. The Seed Savers' Network can be reached at Box 975, Byron Bay, New South Wales 2481, Australia. They have published a well received *Seed Savers' Handbook* that details methods of collecting, cleaning, storing, duplicating and distributing seeds of over 100 crops, in addition to other themes related to seed banking and networking.

Several recent publications have arisen from workshops examining how formal plant breeding can better address the needs and realities faced by small-scale farmers and other local plant germplasm conservators. The first is *Local Knowledge, Global Science, and Plant Genetic Resources* (1994), edited by G.D.

Prain and G.P. Bagalanon and available through Users Perspective with Agricultural Research and Development (UPWARD)/CIP, PCARRD Complex, Los Banos, Philippines. The second compilation is *Using Diversity: Enhancing and Maintaining Genetic Resources On-Farm* (1996), edited by L. Sperling and M. Loevinsohn and available through International Development Research Centre (IDRC), Regional Office for South Asia, 17 Jor Bagh, New Delhi 110 003, India. The third volume along these lines is *Participatory Plant Breeding* (1996), edited by P. Eyzaguirre and M. Iwanaga and available through International Plant Genetic Resources Institute (IPGRI), Via dell Sette Chiese 142, 00145 Rome, Italy.

Finally, information on the bibliography and clearinghouse for intellectual property rights, biodiversity and indigenous peoples' issues, maintained by the Pew Scholars' Conservation Program, can be obtained from Dr Anil Gupta, RM Center for Education Innovation, Vastapur, Ahmedabad 380 015, Gujarat, India.

Appendix B
A model for quantifying the threat of genetic erosion

This model was developed by the International Plant Genetic Resources Institute to 'estimate the threat of genetic erosion that a particular taxon (wild or cultivated) faces in a defined area' [94]. While some of the limitations of the model have already been discussed in Box 5.2, it does represent a way to estimate more quantitatively the range of threats a plant resource may face. Each factor or attribute can be scored either for several taxa in the same region, or for different populations of a plant across several regions [94]. A higher total score indicates a relatively greater degree of imperilment.

Factor		Score
1.	**General**	
1.1	*Taxon distribution:*	
	– rare	10
	– locally common	5
	– widespread or abundant	0
1.2	*Drought:*	
	– known to have occurred in 2 or more consecutive years	10
	– occurring on average one or more times/10 years, but not in consecutive years	5
	– occurring less than once/10 years on average	0
1.3	*Flooding:*	
	– area known to be very flood-prone	10
	– area not known to be flood-prone	0

1.4	*Accidental fires:*	
	– area known to be very prone to fires	10
	– area not known to be fire-prone	0
1.5	*Potential risk from climate change:*	
	– summit areas or low-lying coastal areas	10
2.	**Crop species**	
2.1	*Area under the crop:*	
	– declining rapidly	10
	– increasing or static	0
2.2	*Modern cultivars of the crop:*	
	– available and used by > 70% of farmers	15
	– available and used by 50–70% of farmers	10
	– available and used by < 50% of farmers	5
	– not yet available, but introduction planned	2
	– not available	0
2.3	*Performance of agricultural services:*	
	– very strong, and biased towards modern varieties	10
	– no agricultural services	0
2.4	*Mechanization:*	
	– tractors used by > 30% of farmers	10
	– animal traction used by > 50% of farmers	5
	– manual labor used by > 50% of farmers	0
2.5	*Herbicide and commercial fertilizer use:*	
	– > 50% of farmers	10
	– 25% of farmers	5
	– none	0
2.6	*Farming population:*	
	– declining rapidly	10
	– increasing or static	0
3.	**Wild species**	
3.1	*Extent of wild habitat of target species within study area:*	
	– very restricted (< 5%)	15
	– restricted (5–15%)	10
	– 15–50%	5
	– extensive (> 50%)	0
3.2	*Conservation status of target species:*	
	– not known to occur in any protected area	10
	– known to occur within a protected area, but protection status poor or unknown	5
	– known to occur within a protected area, and protection status good	0

3.3	*Extent of use of wild habitat of target species:*	
	– industrial exploitation	15
	– exploitation by surrounding populations (e.g. fuelwood gathering)	10
	– hunting and gathering by small local communities	2
	– completely protected	0
3.4	*Extent of use of target species:*	
	– industrial exploitation	15
	– exploitation by surrounding populations	10
	– local exploitation	5
	– protected or not used	0
3.5	*Agricultural pressure on wild habitat:*	
	– large-scale cultivation within habitat margins	15
	– subsistence cultivation areas within habitat margins	12
	– land suitable for cultivation, cultivated areas within 3 km of habitat margins	10
	– land suitable for cultivation, cultivated areas within 3–10 km of habitat margins	5
	– land unsuitable for cultivation	0
3.6	*Human population growth rate per year:*	
	– > 3%	10
	– 1–3%	5
	– < 1%	0
3.7	*Availability of agricultural land:*	
	– > 70 ha/km^2 cultivated	10
	– 30–70 ha/km^2 cultivated	5
	– < 30 ha/km^2 cultivated	0
3.8	*Species palatability [to livestock]:*	
	– high	10
	– medium	5
	– low	0
3.9	*Ratio of present livestock density to estimated carrying capacity:*	
	– > 1.0	10
	– 0.5–1.0	5
	– < 0.5	0
3.10	*Average proximity to borehole or other permanent water supply:*	
	– < 10 km	10
	– 10–20 km	5
	– > 20 km	0
3.11	*Distance to major population center:*	
	– < 20 km	10
	– 20–50 km	5
	– > 50 km	0

3.12	*Distance to major road:*	
	– < 10 km	10
	– 10–30 km	5
	– > 30 km	0
3.13	*Proximity to development projects (irrigation scheme, tourism complex, mining site, hydroelectric site, land reclamation scheme):*	
	– < 20 km	10
	– 20–50 km	5
	– > 50 km	0

Reproduced from Guarino (1995) by kind permission of CAB International.

Appendix C Schematic outline for collecting background information on crop varieties

This outline was developed to guide an interdisciplinary field assessment of sweet potato cultivation in Irian Jaya, conducted by Gordon Prain and colleagues [177] at UPWARD, an organization affiliated with the International Potato Center (CIP). Although the fieldwork was targeted towards understanding sweet potato cultivation, the staple crop of highland Irian Jaya, the outline offers a model for community-based research on any assemblage of crops important for *in situ* conservation.

1. Contextual information – defined as information (obtained through observation and interviewing of different informants) on the functioning of the local agricultural system, with particular reference to common practices in sweet potato cultivation.

Location (district, village)
Altitude
Source of information:

- group of farmers (random, local leaders, women, etc.);
- individual farmer (name);
- key informant (name and identity).

Population:

- approximate size, type and name of ethnic group;
- presence of migrants;
- history of the settlement;
- demographic features (migrant males, predominance of an age-group, etc.).

Profile of locality (via transect with member of village, if possible):
- main agroecological zones;
- associated soils;
- associated crops;
- associated livestock;
- associated problems;
- associated opportunities for improvements.

Calendar of activities (associated with rainfall and other climatic factors, if possible):
- crops (especially sweet potato);
- livestock;
- male labor;
- female labor.

2. Crop (sweet potato) cultivation

Varieties:
- approximate number of varieties in the locality;
- why plant so many varieties?
- compare number with the past: more or less?
- why do varieties disappear? Is it important?
- interest in conservation of wide range of varieties;
- who plants most varieties these days?
- important outside sources of information.

Planting material:
- types used and under what circumstances (be specific – tip cuttings, basal cuttings, root combinations);
- sources of material (individual maintenance systems, links with neighbors, etc.);
- form of planting (number of cuttings, how placed in the ground, etc.).

Land preparation, by zone

Planting (use of mounds, on flat, 'pressing under' in garden, etc.).

Cultural practices:
- hilling up;
- weeding;
- use of organic/chemical fertilizer;
- presence of insect pests and diseases;
- use of synthetic and/or 'rustic' pesticides;
- presence of other stresses (waterlogging, drought, rats, etc.).

Ritual practices associated with crop:
- at planting;
- use in rituals;
- use for curing;
- links to women;
- at harvest;
- with food preparation.

Uses of crop parts (e.g. sweet potato tops and roots), by zone:

- estimation of percentages going to different uses;
- estimation of change in percentages during past 10, 20, 30 years;
- how does marketing system work?
- storage, if any;
- processing, if any;
- consumption.

Assessment of overall role, likely changes.

3. Sample plot for variety evaluation

Farmer details:

- name;
- number of family members;
- total area of farm;
- type of tenancy;
- most important crop;
- area of sweet potato.

Plot:

- agroecological zone and cropping system type;
- size;
- other crops.

General comments on crop (sweet potato):

- production problems;
- diversity of varieties: what is the advantage?
- groupings or classifications of varieties: what are the major categories?
- has the farmer encouraged diversity by preserving new types?
- marketing issues, prices, etc. of sweet potato.

4. Comments on varieties planted

Collecting number

Local name(s) of variety:

- known by other names elsewhere?
- widely distributed over locality/other localities?

Physical characteristics (described by farmer)

- root shape and form;
- root skin and flesh color;
- plant type (spreading, compact, where are roots deposited: is it a problem?);
- plant color, texture.

Vegetative period (minimum and normal)

Productivity of roots:

- number of roots;

- size of roots: which important?
- performance in different soils;
- performance over the last few years.

Productivity of tops.

Quality of root:

- floury, watery, sweet, dry, fibrous, other;
- do other family members seek it out on the plate?

Quality of leaves: do people/animals eat them? If so, what is important (succulence, non-hairiness, etc.)?

Effects of climate: special problems or advantages, compared with other crops/varieties?

Effects of insect pests: special problems or advantages, compared with other crops/varieties?

Effects of diseases: special problems or advantages, compared with other crops/varieties?

Reprinted from Prain *et al.* (1995) with kind permission of CAB International.

References

This list includes all references cited in the text of this book. A brief skim of the titles of these references will quickly demonstrate the breadth of topics that are relevant for *in situ* plant conservation. There are plenty of very good materials that we cannot include because of space limitations. Appendix A provides additional suggestions on references we have found particularly useful or comprehensive, as well as addresses for several publishing houses that are leaders in the conservation and community development fields.

1. Alcorn, J.B. (1984) Development policy, forest and peasant farms: reflections on Huastec-managed forests' contributions to commercial production and resource conservation. *Economic Botany* **38**, 389–406.
2. Altieri, M.A. (1995) *Agroecology: The Science of Sustainable Agriculture*, 2nd edn, Westview Press, Boulder, CO.
3. Altieri, M.A. (1987) *Agroecology: The Scientific Basis of Alternative Agriculture*, Westview Press, Boulder, CO.
4. Altieri, M.A. and Montecinos, C. (1993) Conserving crop genetic resources in Latin America through farmers' participation, in *Perspectives on Biodiversity: Case Studies of Genetic Resource Conservation and Development*, (eds C.S. Potter, J.I. Cohen and D. Janczewski), AAAS Press, Washington DC, pp. 45–64.
5. Altieri, M.A. and Hecht, S.B. (1990) *Agroecology and Small Farm Development*, CRC Press, Boca Raton, FL.
6. Altieri, M.A., Anderson, M.K. and Merrick, L.C. (1987) Peasant agriculture and the conservation of crop and wild plant resources. *Conservation Biology*, **1**, 49–58.
7. Altieri, M.A. and Merrick, L.C. (1987) *In situ* conservation of crop genetic resources through maintenance of traditional farming systems. *Economic Botany* **41**(1), 86–96.
8. Altieri, M.A. and Trujillo, J. (1987) The agroecology of corn production in Tlaxcala, México. *Human Ecology* **15**(2), 189–220.
9. Amador, M.F. (1980) Comportamiento de tres especies (maiz, frijol, calabaza) en policultivos en La Chontalpa, Tabasco, México. Tesis Profesional, CSAT, Cardenas, Tabasco, México.
10. Anderson, E. (1954) *Plants, Man, and Life*, Andrew Melrose, London.
11. Anderson, M.K. and Nabhan, G.P. (1991) Gardeners in Eden, *Wilderness* **55**(194), 27–30.
12. Andrade Aguilar, J.A. and Hernandez Xolocotzi, E. (1991) Diversity of common beans (*Phaseolus vulgaris* Fabaceae) and conditions of production in Aguascalientes, México. *Economic Botany* **45**(3), 339–344.

13. Anon. (1995) The computer-assisted resource inventory system (CAMRIS) an inexpensive, user-friendly geographic information system. Unpublished information sheet, RARE Center for Tropical Conservation, Philadelphia, PA.
14. Atay, S. (1996) Progress report on bulb propagation, Turkey. Report for WWF International. Turkish Society for the Protection of Nature, Istanbul.
15. Axelrod, D.I. (1970) *Age and Origin of Sonoran Desert Vegetation*, California Academy of Sciences, San Francisco.
16. Balick, M.J. and Cox, P.A. (1996) *Plants, People and Culture: The Science of Ethnobotany*, Scientific American Library, New York.
17. Balick, M.J., Arvigo, R. and Romero, L. (1994) The development of an ethnobiomedical forest reserve in Belize: its role in the preservation of biological and cultural diversity. *Conservation Biology* **8**(1), 316–317.
18. Barbour, M.G., Burk, J.H. and Pitts, W.D. (1987) *Terrestrial Plant Ecology*, Benjamin Cummings Publishing, Menlo Park, CA.
19. Batisse, M. (1982) The biosphere reserve: a tool for environmental conservation and management. *Environmental Conservation* **9**(2), 101–111.
20. Bedasse, J. and Stewart, N. (1996) The Maroons of Jamaica: one with mother earth, in *Traditional Peoples and Biodiversity Conservation in Large Tropical Landscapes* (eds K.H. Redford and J.A. Mansour), America Verde Publications, The Nature Conservancy, Latin America and Caribbean Division, Arlington, VA, pp. 57–74.
21. Bellon, M.R. and Brush, S.B. (1994) Keepers of maize in Chiapas, Mexico. *Economic Botany* **48**(2), 196–209.
22. Bennett, P., Johnson, R.R. and Kunzman, M.R. (1987) Cactus collection factors of interest to resource managers, in *Conservation and Management of Endangered Plants* (ed. T.S. Elias), California Native Plant Society, Sacramento, CA, pp. 215–224.
23. Benz, B.F., Sanchez-Velasquez, L.R. and Santana Michel, F.J. (1990) Ecology and ethnobotany of *Zea diploperennis*: preliminary investigations. *Maydica* **35**, 85–98.
24. Berg, T. (1996a) The compatability of grassroots breeding and modern farming, in *Participatory Plant Breeding* (eds P. Eyzaguirre and M. Iwanaga), IPGRI, Rome.
25. Berg, T. (1996b) Devolution of plant breeding, in *Using Diversity: Enhancing and Maintaining Genetic Resources On-Farm* (eds L. Sperling and M. Loevinsohn), International Development Research Centre, New Delhi, India.
26. Berlin, B. (1992) *Ethnobiological Classification: Principles of Categorization of Plants and Animals in Traditional Societies*, Princeton University Press, Princeton, NJ.
27. Berlin, B., Breedlove, D.E. and Raven, P.H. (1974) *Principles of Tzeltal Plant Classification*, Academic Press, New York and London.
28. Borgtoft-Pedersen, H. and Balsev, H. (1990) *Ecuadorean Palms for Agroforestry*, AAU Reports 23, Aarhus University Press, Aarhus, Denmark.
29. Borrini-Feyerabend, G. and Buchan, D. (1996) *Beyond Fences: Seeking Social Sustainability in Conservation*, Vol. II, IUCN, Gland, Switzerland.
30. Brand, J., Snow, B.J., Nabhan, G.P. and Truswell, A. (1990) Plasma glucose and insulin responses to traditional Pima Indian meals. *American Journal of Clinical Nutrition* **51**, 416–420.
31. Brandon, K. and Wells, M. (1992) Planning for people and parks: design dilemmas. *World Development* **20**(4), 557–570.
32. Bright, C. (1997) Tracking the ecology of climate change, in *State of the World 1997* (ed. L. Starke), Worldwatch Institute, Washington DC, pp. 78–94.
33. Bronstein, J.L., Gouyon, P.H., Gliddon, C., Kjeliberg, F. and Michaloud, G. (1990) The ecological consequences of flowering asynchrony in monoecious figs: a simulation study. *Ecology* **71**(6), 2145–2146.

34. Brookfield, H. and Padoch, C. (1994) Appreciating agrodiversity: a look at the dynamism and diversity of indigenous farming practices. *Environment* **36**(5), 7–11, 37–45.
35. Brower, J.E. and Zar, J.H. (1977) *Field and Laboratory Methods for General Ecology*, Wm C. Brown Publishers, Dubuque, IA.
36. Bruce, J.W. (1990) *Community Forestry: Rapid Appraisal of Tree and Land Tenure*, Community Forestry Note No. 5, FAO/UN, Rome.
37. Brush, S.B. (1996) Whose knowledge, whose genes, whose rights? in *Valuing Local Knowledge: Indigenous People and Intellectual Property Rights* (eds S.B. Brush and D. Stabinsky), Island Press, Washington DC, pp. 41–67.
38. Brush, S.B. (1995) *In situ* conservation of landraces in centers of crop diversity. *Crop Science* **35**, 346–354.
39. Brush, S.B. (1991) A farmer-based approach to conserving crop germplasm. *Economic Botany* **45**(2), 153–165.
40. Brush, S.B., Kesseli, R., Ortega, R. *et al.* (1995) Potato diversity in the Andean center of crop domestication. *Conservation Biology* **9**(5), 1189–1198.
41. Brush, S.B., Taylor, J.E. and Bellon, M. (1992) Technology adoption and biological diversity in Andean potato agriculture. *Journal of Development Economics* **39**, 365–387.
42. Buchmann, S.L. and Nabhan, G.P. (1996) *The Forgotten Pollinators*, Island Press, Washington DC.
43. Bunch, R. and Lopez, G. (1995) *Soil Recuperation in Central America: Sustaining Innovation After Intervention*, IIED Gatekeepers Series, No. 55, International Institute for Environment and Development, London.
44. Center for Plant Conservation (1990) *Genetic Sampling Guidelines for Conservation Collecting of Endangered Plants*, Center for Plant Conservation, Jamaica Plain, MA.
45. Chang, J.H. (1977) Tropical agriculture: crop diversity and crop yields. *Economic Geography* **53**, 241–254.
46. Chapin, M. (1994) Recapturing the old ways: traditional knowledge and western science among the Kuna Indians of Panama, in *Cultural Expression and Grassroots Development* (ed. C.D. Kleymeyer), Lynne Rienner Publishing, Boulder, CO, pp. 83–102.
47. Chernela, J. (1987) Endangered ideologies: Tukano fishing taboos. *Cultural Survival Quarterly* **11**, 50–52.
48. Chinwuba Obi, S.N. (1988) Rights in economic trees, in *Whose Trees? Proprietary Dimensions of Forestry* (eds L. Fortmann and J.W. Bruce), Westview Press, Boulder, CO, pp. 34–38.
49. Clawson, D.L. (1985) Harvest security and intraspecific diversity in traditional tropical agriculture. *Economic Botany* **39**, 56–67.
50. Clay, J.W. (1996) *Generating Income and Conserving Resources: 20 Lessons from the Field*, World Wildlife Fund, Washington DC.
51. Clay, J.W. (1988) *Indigenous Peoples and Tropical Forests: Models of Land Use and Management from Latin America*, Cultural Survival Report No. 27, Cultural Survival, Inc., Cambridge, MA.
52. Colchester, M. (1994) *Salvaging Nature: Indigenous Peoples, Protected Areas and Biodiversity Conservation*, Discussion Paper 55, United Nations Research Institute for Social Development, Geneva, Switzerland.
53. Conklin, H.C. (1956) *Huananoo Agriculture: a report of an integral system of shifting cultivation in the Philippines*, Forestry Development Paper No. 12, FAO/UN, Rome.
54. Cordell, J. (1993) Boundaries and bloodlines: tenure of indigenous homelands and protected areas, in *The Law of the Mother: Protecting Indigenous Peoples in Protected Areas* (ed. E. Kemf), Earthscan Publications, London.

55. Craven, I. and Wardoyo, W. (1993) Gardens in the forest, in *The Law of the Mother: Protecting Indigenous Peoples in Protected Areas* (ed. E. Kemp), Earthscan Publications, London.
56. Cronk, Q. and Fuller, J. (1995) *Plant Invaders*, People and Plants Conservation Manual 2, Chapman & Hall, London.
57. Cronon, W. (1983) *Changes in the Land: Indians, Colonists, and the Ecology of New England*, Hill and Wang, New York.
58. Cunningham, A.B. *People and Wild Plant Use* People and Plants Conservation Manual 4, Chapman & Hall, London. (In prep.)
59. Cunningham, A.B. (1996) Professional ethics and ethnobotanical research, in *Selected Guidelines for Ethnobotanical Research: A Field Manual* (ed. M.N. Alexiades), The New York Botanical Garden, Bronx, NY, pp. 19–52.
60. Cunningham, A.B. (1993) *African Medicinal Plants: Setting Priorities at the Interface Between Conservation and Primary Healthcare*, People and Plants Working Paper 1, UNESCO, Paris.
61. Dasmann, R. (1984) The relationship between protected areas and indigenous peoples, in *National Parks, Conservation and Development* (eds J.A. McNeely and K.R. Miller), Smithsonian Press, Washington DC.
62. Davis-Case, D. (1990) *The Community's Toolbox: The Idea, Methods and Tools for Participatory Assessment, Monitoring and Evaluation in Community Forestry*, Community Forestry Field Manual 2, FAO/UN, Rome.
63. Deb Roy, S. and Jackson, P. (1993) Mayhem in Manas: the threats to India's wildlife reserves, in *The Law of the Mother: Protecting Indigenous Peoples in Protected Areas* (ed. E. Kemf), Earthscan Publications, London.
64. Doebley, J. (1990) Molecular evidence for gene flow among *Zea* species. *BioScience* **40**(6), 443–448.
65. Doebley, J.F. and Nabhan, G.P. (1989) Further evidence regarding gene flow between maize and teosinte. *Maize Genetics Cooperation Newsletter* **63**, 107–108.
66. Dove, M.R. (1993) The responses of Dayak and bearded pig to mast-fruiting in Kalimantan: an analysis of nature–culture analogies, in *Tropical Forests, People and Food: Biocultural Interactions and Applications to Development* (eds C.M. Hladik, A. Hladik, O.F. Linares *et al.*), UNESCO, Paris, pp. 113–126.
67. Durning, A.T. (1992) *Guardians of the Land: Indigenous Peoples and the Health of the Earth*, Worldwatch Paper No. 12, Worldwatch Institute, Washington DC.
68. Dvorak, W.S. and Donahue, J.K. (1992) *CAMCORE Cooperative Research Review 1980–1992*, Department of Forestry, North Carolina State University, Raleigh, NC.
69. Dvorak, W.S. and Laarman, J.G. (1986) Conserving the genes of tropical conifers. *Journal of Forestry* **84**(1), 43–45.
70. Ecker, L.S. (1990) Population enhancement of a rare Arizona cactus, *Mammilaria thornberi* Orcutt (Cactaceae). Master's Thesis, Department of Botany, Arizona State University, Tempe.
71. Ellstrand, N.C. (1992) Gene flow among seed plant populations. *New Forests* **6**, 241–256.
72. Endler, J.A. (1977) *Geographic Variation, Speciation, and Clines*, Princeton University Press, Princeton, NJ.
73. English, J., Tiffen, M. and Mortimore, M. (1994) *Land Resource Management in Machakos District, Kenya 1930–1990*, Environment Paper No. 5, The World Bank, Washington DC.
74. Falk, D.A. and Holsinger, K. (1991) *Genetics and Conservation of Rare Plants*, Oxford University Press, New York.

75. Falk, D.A., Millar, C.I. and Olwell, M. (1996) *Restoring Diversity: Strategies for the Reintroduction of Endangered Plants*, Island Press, Washington DC.
76. Felger, R.S., Nabhan, G.P. and Bye, R.A. Jr (1997) *The Northern Sierra Madre Occidental as a Center of Megadiversity*, IUCN (Botanical Garden Conservation Sec.)/WWF, Godalming, Surrey, UK.
77. Fisher, L. and Bunch, R. (1996) Challenges in promoting forest patches in rural development efforts, in *Forest Patches in Tropical Landscapes* (eds J. Schelhas and R. Greenberg), Island Press, Washington DC, pp. 300–326.
78. Food and Agriculture Organization of the United Nations (1996) *The State of the World's Plant Genetic Resources for Food and Agriculture* (background documentation prepared for the International Technical Conference on Plant Genetic Resources, Leipzig, Germany, 17–23 June 1996), FAO/UN, Rome.
79. Fowler, C. and Mooney, P.R. (1990) *Shattering: Food, Politics, and the Loss of Genetic Diversity*, University of Arizona Press, Tucson, AZ.
80. Francis, C.A. (1986) *Multiple Cropping Systems*, MacMillan, New York.
81. Franzen, H., Ay, P., Begemann, F. *et al.* (1996) Variety improvement in the informal sector: aspects of a new strategy, in *Participatory Plant Breeding* (eds P. Eyzaguirre and M. Iwanaga), IPGRI, Rome.
82. Freudenberger, K.S. and Gueye, B. (1990) *RRA Notes to Accompany Introductory Training Module*, International Institute for Environment and Development, London.
83. Gadgil, M. (1992) Conserving biodiversity as if people matter: a case study from India. *Ambio* **21**(3), 266–270.
84. Gentry, A.H. (1988) Tree species richness of upper Amazonian forests. *Proceedings of the National Academy of Sciences* **85**, 156–159.
85. Gentry, A.H. (1986) Endemism in tropical versus temperate plant communities, in *Conservation Biology: The Science of Scarcity and Diversity* (ed. M.E. Soule), Sinauer Associates, Sunderland, MA, pp. 153–181.
86. Giannini, I.V. (1996) The Xikrin do Catetë Indigenous Area, in *Traditional Peoples and Biodiversity Conservation in Large Tropical Landscapes* (eds K.H. Redford and J.A. Mansour), America Verde Publications, The Nature Conservancy, Latin America and Caribbean Division, Arlington, VA, pp. 115–136.
87. Gilpin, M.E. and Soule, M.E. (1986) Minimum viable populations: processes of species extinctions, in *Conservation Biology: The Science of Scarcity and Diversity* (ed. M.E. Soule), Sinauer Associates, Sunderland, MA, pp. 19–34.
88. Gliessman, S.R. (1990) The ecology and management of traditional farming systems, in *Agroecology and Small Farm Development* (eds M.A. Altieri and S.B. Hecht), CRC Press, Boca Raton, FL.
89. Gliessman, S.R. (1983) Allelopathic interactions in crop–weed mixtures: applications for weed management. *Journal of Chemical Ecology* **9**, 991.
90. Gliessman, S.R. (1982) Nitrogen distribution in several traditional agro-ecosystems in the humid tropical lowlands of south-eastern Mexico. *Plant Soil* **67**, 105.
91. Gómez-Pompa, A., Salvador, F.J. and Sosa, V. (1987) The 'pet kot': a man-made tropical forest of the Maya. *Interciencia* **12**(1), 10–15.
92. Gonzalez, M.M. and Bosland, P.W. (1991) Strategies for stemming the genetic erosion of *Capsicum* germplasm in the Americas. *Diversity* **7**, 52–53.
93. Gonzalez, R.M. (1995) KBS, GIS and documenting indigenous knowledge. *Indigenous Knowledge and Development Monitor* **3**(1),5–7.
94. Guarino, L. (1995) Assessing the threat of genetic erosion, in *Collecting Plant Diversity: Technical Guidelines* (eds L. Guarino, V.R. Rao and R. Reid), CAB International, Wallingford, Oxon, UK, pp. 67–74.

95. Guarino, L., Rao, V.R. and Reid, R. (eds) (1995) *Collecting Plant Diversity: Technical Guidelines*, CAB International, Wallingford, Oxon, UK.
96. Gupta, M.P., Jones, A., Solís, P. and Correa, A.M.D. (1992) Ethnopharmacognostic study of Kuna Yala, in *Sustainable Harvest and Marketing of Rain Forest Products* (eds M. Plotkin and L. Famolare), Island Press and Conservation International, Washington DC, pp. 201–211.
97. Halffter, G. (1985) Biosphere reserves: conservation of nature for man. *Parks* **10**(3), 15–-18.
98. Hamilton, A. (1997), personal communication to JT.
99. Hamilton, A. (1995) Threats to plants: an analysis of *Centres of Plant Diversity*, World Wide Fund for Nature, Godalming, Surrey, UK.
100. Harlan, J.R. (1989) Wild grass seed harvesting in the Sahara and sub-Sahara of Africa, in *Foraging and Farming: The Evolution of Plant Exploitation* (eds D.R. Harris and G.C. Hillman), Unwin-Hyman, London.
101. Harroy, J.P. (1968) The development of the natioal park movement, in *The Canadian National Parks: Today and Tomorrow* (eds J.C. Nelson and R.C. Scace), Studies in Land Use History and Lancscape Change, National Parks Series No. 3, University of Calgary, Alberta, Canada, pp. 17–34.
102. Hecht, S.B. (1987) The evolution of agroecological thought, in *Agroecology: The Scientific Basis of Alternative Agriculture* (ed. M.A. Altieri), Westview Press, Boulder, CO, pp. 1–14.
103. Hills, J. and Hine (1985) Seri maps. Unpublished manuscript presented to American Association of Geographers, in possession of authors, Tucson, Arizona.
104. Hiraoka, M. (1992) Caboclo and Ribereño resource management in Amazonia: a review, in *Conservation of Neotropical Forests: Working from Traditional Resource Use* (eds K.H. Redford and C. Padoch), Columbia University Press, New York, pp. 134–157.
105. Hodgson, W. (1993), personal communication to GN.
106. Hoyt, E. (1990) Wild relatives. *Wilderness* (Summer), 45–54.
107. Hoyt, E. (1988) *Conserving the Wild Relatives of Crops*, IBPGR, Rome.
108. Hunn, E. (1975) A measure of the degree of correspondence of folk to scientific biological classification. *American Ethnologist* **2**(2), 309–327.
109. Iltis, H.H. (1988) Serendipity in the evolution of biodiversity: what good are weedy tomatoes?, in *Biodiversity* (ed. E.O. Wilson), National Academy Press, Washington DC, pp. 98–105.
110. Iltis, H.H. (1974) Freezing the genetic landscape – the preservation of diversity in cultivated plants as an urgent social responsibility of the plant geneticist and plant taxonomist. *Maize Genetics Cooperation Newsletter* **48**, 199–200.
111. Ingram, G.B. and Williams, J.T. (1984) *In situ* conservation of wild relatives of crops, in *Crop Genetic Resources: Conservation and Evaluation* (eds J.H.W. Holden and J.T. Williams), IBPGR, Rome, pp. 163–179.
112. International Board for Plant Genetic Resources (1992) Rare durian rescued from extinction. *Geneflow*, IBPGR, Rome.
113. IUCN Commission on National Parks and Protected Areas (1984) Categories, objectives and criteria for protected areas, in *National Parks, Conservation and Development* (eds J.A. McNeely and K.R. Miller), Smithsonian Press, Washington DC, pp. 47–53.
114. Jacobsohn, M. (1993) Conservation and a Himba community in western Kaokoland, in *Voices from Africa: Local Perspectives on Conservation* (eds D. Lewis and N. Carter), World Wildlife Fund, Washington DC, pp. 99–112.

115. Kaus, A. (1993) Environmental perceptions and social relations in the Mapimí Biosphere Reserve. *Conservation Biology* **7**(2), 398–406.
116. Kearns, C.A. and Inouye, D.W. (1993) *Techniques for Pollination Biologists*, University Press of Colorado, Niwot, CO.
117. Khedkar, R. (1996) The Academy of Development Sciences Rice Project: need for decentralized community genebanks to strengthen on-farm conservation, in *Using Diversity: Enhancing and Maintaining Genetic Resources On-Farm* (eds L. Sperling and M. Loevinsohn), International Development Research Centre, New Delhi, India.
118. Khon Kaen University (1985) *Proceedings of the 1985 International Conference of Rapid Rural Appraisal*, Khon Kaen University, Thailand.
119. Kiambi, K. and Opole, M. (1992) Promoting traditional trees and food plants in Kenya, in *Growing Diversity: Genetic Resources and Local Food Security* (eds D. Cooper, R. Vellve and H. Hobbelink), Intermediate Technology Publications, London, pp. 53–68.
120. Lahm, S. (1993) Utilization of forest resources and local variation of wildlife populations in northeastern Gabon, in *Tropical Forests, People and Food: Biocultural Interactions and Applications to Development* (eds C.M. Hladik, A. Hladik, O.F. Linares *et al.*), UNESCO, Paris, pp. 213–226.
121. Laurent, P.-J. and Mathieu, P. (1994) Authority and conflict in management of natural resources – a story about trees and immigrants in southern Burkina Faso. *Forests, Trees and People Newsletter* **25** (October), 37–44.
122. Lebbie, A.R. and Freudenberger, M.S. (1996) Sacred groves in Africa: forest patches in transition, in *Forest Patches in Tropical Landscapes* (eds J. Schelhas and R. Greenberg), Island Press, Washington DC, pp. 300–326.
123. Lebbie, A.R. and Guries, R.P. (1995) Ethnobotanical value and conservation of sacred groves of the Kpaa Mende in Sierra Leone. *Economic Botany* **49**(3), 297–308.
124. Leon Zuniga, L. (1994) Hear ye, hear ye! *Forests, Trees and People Newsletter* **24** (June), 42–43.
125. Letourneau, D.K. (1990) Two examples of natural enemy augmentation: a consequence of crop diversification, in *Agroecology: Researching the Ecological Basis of Sustainable Agriculture* (ed. S.R. Gliessman), Springer-Verlag, Berlin, pp. 11–29.
126. Lewis, D. (1990) Conflict of interests. *Geographical Magazine* (December), 18–22.
127. Linares, O.F. (1976) 'Garden hunting' in the American tropics. *Human Ecology* **4**(4), 331–349.
128. Litsinger, J.A. (1993) A farming systems approach to insect pest management for upland and lowland rice farmers in tropical Asia, in *Crop Protection Strategies for Subsistence Farmers* (ed. M.A. Altieri), Westview Press, Boulder, CO.
129. Loaiza-Figueroa, F., Ritland, K., Laborde-Cancino, J.A. and Tanksley, S.D. (1989) Patterns of genetic variation of the genus *Capsicum* (Solanaceae) in México. *Plant Systematics and Evolution* **165**, 159–188.
130. McAuliffe, J.R. (1990) A rapid survey method for the estimation of density and cover in desert plant communities. *Journal of Vegetation Science* **1**, 653–656.
131. McNeely, J.A. and Miller, K.R. (1984) *National Parks, Conservation and Development*, Smithsonian Press, Washington DC.
132. Margoluis, R. and Salafsky, N. (1998) *Measures of Success: Designing, Managing and Monitoring Conservation and Development Projects*, Island Press, Washington DC.
133. Martin, G.J. (1995) *Ethnobotany*, People and Plants Conservation Manual 1, Chapman & Hall, London.
134. Martin, G.J., Beaman, J.H. Beaman, R. *et al.* (in preparation) Efficacy of community based botanical inventories.

135. Maxted, N., van Slageren, M.W. and Rihan, J.R. (1995) Ecogeographic surveys, in *Collecting Plant Diversity: Technical Guidelines* (eds L. Guarino, V.R. Rao and R. Reid), CAB International, Wallingford, Oxon, UK, pp. 255–285.
136. May, R.M., Lawton, J.H. and Stork, N.E. (1995) Assessing extinction rates, in *Extinction Rates* (eds J.H. Lawton and R.M. May), Oxford University Press, Oxford, UK, pp. 1–24.
137. May, R.M. and Lyle, A.M. (1987) Living Latin binomials: conservation biology. *Nature* **326**, 642–643.
138. Meffe, G.K. and Carroll, C.R. (1994) *Principles of Conservation Biology*, Sinauer Associates, Sunderland, MA.
139. Menges, E.S. (1991) The application of minimum viable population theory to plants, in *Genetics and Conservation of Rare Plants* (eds D.A. Falk and K. Holsinger), Oxford University Press, New York, pp. 31–61.
140. Menges, E.S. and Gordon, D.R. (1996) Three levels of monitoring intensity for rare plant species. *Natural Areas Journal* **16**(3), 227–237.
141. Millar, C.I. and Libby, W.J. (1991) Strategies for conserving clinal, ecotypic and disjunct population diversity in widespread species, in *Genetics and Conservation of Rare Plants* (eds D.A. Falk and K. Holsinger), Oxford University Press, New York, pp. 149–170.
142. Nabhan, G.P. (1994) Proximate and ultimate threats to endangered species. *Conservation Biology* **8**(4), 928–929.
143. Nabhan, G.P. (1992) Native plant products from the arid neotropical species: assessing benefits to cultural, environmental and genetic diversity, in *Sustainable Harvest and Marketing of Rain Forest Products* (eds M. Plotkin and L. Famolare), Island Press and Conservation International, Washington DC, pp. 137–140.
144. Nabhan, G.P. (1990a) *Wild* Phaseolus *Ecogeography in the Sierra Madre Occidental, Mexico: Aerographic Techniques for Targeting and Conserving Species Diversity*, Systematic and Ecogeographical Studies on Crop Genepools 5, IBPGR and FAO/UN, Rome.
145. Nabhan, G.P. (1990b) *Cultures of Habitat*, Counterpoint Press, Washington, D.C.
146. Nabhan, G.P. (1989) *Enduring Seeds: Native American Agriculture and Wild Plant Conservation*, North Point Press, San Francisco, CA.
147. Nabhan, G.P. (1987) Nurse plant ecology of threatened desert plants, in *Conservation and Management of Endangered Plants* (ed. T.S. Elias), California Native Plant Society, Sacramento, CA, pp. 377–384.
148. Nabhan, G.P. (1985a) *Gathering the Desert*, University of Arizona Press, Tucson, AZ.
149. Nabhan, G.P. (1985b) Native crop diversity in Aridoamerica: conservation of regional gene pools. *Economic Botany* **39**(4), 387–399.
150. Nabhan, G.P. (1982) *The Desert Smells Like Rain*, University of Arizona Press, Tucson, AZ.
151. Nabhan, G.P. and Suzan, H. (1994) Boundary effects on endangered cacti and their nurse plants in and near a Sonoran Desert biosphere reserve, in *Ironwood: An Ecological and Cultural Keystone of the Sonoran Desert* (eds G.P. Nabhan and J.L. Carr), Occasional Papers in Conservation Biology, No. 1, Conservation International, Washington DC, pp. 55–69.
152. Nabhan, G.P. (ed.) (1998) *Passing on a Sense of Place and Ecological Knowledge Between Generations: A Primer*. Arizona-Sonora Desert Museum, Tucson, AZ.
153. Nabhan, G.P., House, D., Suzan, A. *et al.* (1991) Conservation and use of rare plants by traditional cultures of the US/Mexico borderlands,, in *Biodiversity: Culture,*

Conservation and Ecodevelopment (eds M.L. Oldfield and J.B. Alcorn), Westview Press, Boulder, CO, pp. 127–146.
154. Nabhan, G.P., Slater, M. and Yarger, L. (1990) New crops for marginal peoples in marginal lands: wild chiles as a case study, in *Agroecology and Small Farm Development* (eds M.A. Altieri and S.B. Hecht), CRC Press, Boca Raton, FL, pp. 19–26.
155. Nabhan, G.P. and Felger, R.S. (1985) Wild relatives of crops: their direct uses as food, in *Plants for Arid Lands* (ed. G.E. Wickens), George Allen and Unwin, London, pp. 19–35.
156. Nash, R. (1967) *Wilderness and the American Mind*, Yale University Press, New Haven, CT.
157. Oates, J.F. (1995) The dangers of conservation by rural development – a case study from the forests of Nigeria. *Oryx* **29**(2), 115–122.
158. Odour Noah, E., Asamba, I., Ford, R. *et al.* (1992) *Implementing PRA: A Handbook to Facilitate Participatory Rural Appraisal*, Program for International Development, Clark University, Worcester, MA.
159. Oldfield, M.L. (1984) *The Value of Conserving Genetic Resources*, Sinauer Associates, Sunderland, MA.
160. Oldfield, M.L. and Alcorn, J.B. (eds) (1991) *Biodiversity: Culture, Conservation and Ecodevelopment*, Westview Press, Boulder, CO.
161. Osborn, T. (1995) *Participatory Agricultural Extension: Experiences from West Africa*, IIED Gatekeeper Series, No. 48, International Institute for Environment and Development, London.
162. Pelkey, N. (1996) PRA & RRA: A dangerous paradigm? *ETFRN News* **18/96**, 18–20.
163. Peluso, N.L. (1992) The ironwood problem: (mis)management and development of an extractive rainforest product. *Conservation Biology* **6**, 210–219.
164. Peluso, N.L. (1993) *Rich Forests, Poor People*, University of California Press, Berkeley, CA.
165. Pennington, C.W. (1969) *The Tepehuan of Chihuahua: Their Material Culture*, University of Utah Press, Salt Lake City, UT.
166. Peters, C.M. (1996) *The Ecology and Management of Non-timber Forest Resources*, World Bank Technical Paper No. 322, The World Bank, Washington DC.
167. Peters, C.M. (1994) *Sustainable Harvest of Non-timber Plant Resources in Tropical Moist Forest: An Ecological Primer*, USAID Biodiversity Support Program, Washington DC.
168. Petit, S. and Pors, L. (1996) Survey of columnar cacti and carrying capacity for nectar-feeding bats on Curaçao. *Conservation Biology* **10**(3), 769–775.
169. Phillips, O., Gentry, A.H., Reynel, C. *et al.* (1994) Quantitative ethnobotany and Amazonian conservation. *Conservation Biology* **8**(1), 225–248.
170. Pimbert, M.P. and Pretty, J.N. (1995) *Parks, People and Professionals: Putting 'Participation' into Protected Area Management*, Discussion Paper 57, United Nations Research Institute for Social Development, Geneva, Switzerland.
171. Pinedo-Vasquez, M. and Padoch, C. (1993) Community and governmental experiences protecting biodiversity in the lowland Peruvian Amazon, in *Perspectives on Biodiversity: Case Studies of Genetic Resource Conservation and Development* (eds C.S. Potter, J.I. Cohen and D. Janczewski), AAAS Press, Washington DC, pp. 199–211.
172. Pinedo-Vasquez, M., Zarin, D., Jipp, P. and Chota-Inuma, J. (1990) Use-values of tree species in a communal forest reserve in northeast Peru. *Conservation Biology* **4**, 405–416.

173. Plotkin, M.J. (1994) *Tales of a Shaman's Apprentice*, Viking Press, New York.
174. Poffenberger, M. (1994) The resurgence of community forest management in eastern India, in *Natural Connections: Perspectives in Community-based Conservation* (eds D. Western, R.M. Wright and S.C. Strum), Island Press, Washington DC, pp. 53–79.
175. Poffenberger, M., McGean, B., Ravindranath, N.H. and Gadgil, M. (1992) *Field Methods Manual, Volume I: Diagnostic Tools for Supporting Joint Forest Management Systems*, Joint Forest Management Support Program, Society for Promotion of Wastelands Development, New Delhi, India.
176. Poole, P. (1995) *Indigenous Peoples, Mapping and Biodiversity Conservation: An Analysis of Current Activities and Opportunities for Applying Geomatics Technologies*, Biodiversity Support Program, Washington DC.
177. Prain, G.D. Gin-Mok, I., Sawor, T. *et al.* (1995) Interdisciplinary collecting of *Ipomoea batatas* germplasm and associated indigenous knowledge in Irian Jaya, in *Collecting Plant Diversity: Technical Guidelines* (eds L. Guarino, V.R. Rao and R. Reid), CAB International, Wallingford, Oxon, UK.
178. Prance, G.T., Balee, W., Boom, B.M. and Carneiro, R.L. (1987) Quantitative ethnobotany and the case for conservation in Amazonia. *Conservation Biology* **1**, 296–310.
179. Prescott-Allen, R. and Prescott-Allen, C. (1984) Park your genes: protected areas as *in situ* genebanks for the maintenance of wild genetic resources, in *Proceedings from the 1982 Bali World Congress on National Parks* (eds J.A. McNeely and K.R. Miller), Smithsonian Institution Press, Washington DC.
180. Primack, R.B. (1993) *Essentials of Conservation Biology*, Sinauer Associates, Sunderland, MA.
181. Quisumbing, A.R., Brown, L.R., Sims-Feldstein, H. *et al.* (1995) *Women: The Key to Food Security*, Food Policy Reports series, IFPRI, Washington DC.
182. Rabinovitch-Vin, A. (1991) Continuous human use as a tool for species richness in protected areas of Israel, in *Resident People and National Parks* (eds P.C. West and S.R. Brechin), University of Arizona Press, Tucson, AZ, pp. 95–100.
183. Rabinowitz, D., Linder, C.R., Ortega, R. *et al.* (1990) High levels of interspecific hybridization between *Solanum sparsipilum* and *S. stentomum* in experimental plots in the Andes. *American Potato Journal* **67**(2), 73–81.
184. Rabinowitz, D., Cairns, S. and Dillon, T. (1986) Seven forms of rarity and their frequency in the flora of the British Isles, in *Conservation Biology: The Science of Scarcity and Diversity* (ed. M.E. Soule), Sinauer Associates, Sunderland, MA, pp. 182–220.
185. Rapoport, E.H. (1975) *Areografía: Estrategias Geográficas de las Especies*, Fondo de Cultura Economica, México DF, México.
186. Raven, P.H. (1994), personal comment to MH.
187. Rea, A.M. (1983) *Once a River*, University of Arizona Press, Tucson, AZ.
188. Rea, A.M. (1997) *The Desert's Green Edge: A Pima Ethnobotany*, University of Arizona Press, Tucson.
189. Redford, K.H. (1993) Hunting in neotropical forests: a subsidy from nature, in *Tropical Forests, People and Food: Biocultural Interactions and Applications to Development* (eds C.M. Hladik, A. Hladik, O.F. Linares *et al.*), UNESCO, Paris, pp. 227–248.
190. Reichhardt, K.L., Mellink, E., Nabhan, G.P. and Rea, A.M. (1994) Habitat heterogeneity and biodiversity associated with indigenous agriculture in the Sonoran desert. *Ethnoecológica* **2**(3), 21–34.
191. Riley, K.W. (1996) Decentralized breeding and selection: tool to link diversity and development, in *Using Diversity: Enhancing and Maintaining Genetic Resources On-*

Farm (eds L. Sperling and M. Loevinsohn), International Development Research Centre, New Delhi, India.

192. Robertson, S.A. (1992) NMK and WWF-I: Working with the Elders to Protect the Sacred Kaya Forests of Coastal Kenya. Unpublished report, WWF International, Gland, Switzerland.
193. Rocheleau, D., Ross, L., Morrobel, J. and Hernandez, R. (1996) *Forests, Gardens and Tree Farms: Gender, Class and Community at Work in the Landscapes of Zambrana-Chacuey*, ECOGEN Case Studies Series, Clark University, Worcester, MA.
194. Rocheleau, D., Thomas-Slayter, B. and Edmunds, D. (1995) Gendered resource mapping: focusing on women's spaces in the landscape. *Cultural Survival Quarterly* **18**(4), 62–68.
195. Rocheleau, D., Weber, F. and Field-Juma, A. (1989) *Agroforestry in Dryland Africa*, International Council for Research in Agroforestry, Nairobi, Kenya.
196. Roos, E.E. (1986) Precepts of successful seed storage, in *Physiology of Seed Deterioration*, Special Publication No. 11, Crop Science Society of America, Madison, WI.
197. Rosenberg, J. (1996) Application of a model for evaluating biosphere reserves: a case study of the proposed Tiburon–San Esteban Biosphere Reserve in the Gulf of California. Unpublished manuscript, University of Arizona, Tucson, AZ.
198. Ruddell, E.D. and Beingolea, J. (1995) Towards farmer scientists. *ILEIA Newsletter* **11**(1), 16–17.
199. Rwangyezi, S. and Woomer, P. (1995) Promoting environmental awareness through performance education. *Nature and Resources* **31**(4), 34–39.
200. Rzedowski, J. (1991) Diversidad y orígenes de la flora fanerogámica de México. *Acta Botan. Mexico* **14**, 3–21.
201. Salazar, R. (1992) Community plant genetic resources management: experiences in Southeast Asia, in *Growing Diversity: Genetic Resources and Local Food Security* (eds D. Cooper, R. Vellve and H. Hobbelink), Intermediate Technology Publications, London, pp. 17–30.
202. Salick, J., Cellinese, N. and Knapp, S. (1997) Indigenous diversity of cassava: generation, maintenance, use and loss among the Amuesha, Peruvian upper Amazon. *Economic Botany* **51**(1), 6–19.
203. Salick, J. and Merrick, L.C. (1990) Use and maintenance of genetic resources: crops and their wild relatives, in *Agroecology* (eds P. Rosset, C.R. Carroll and J.H. Vandermeer), McGraw-Hill, New York, pp. 517–548.
204. Santos, M.M. (1995) Farmer's rights and access to indigenous seeds: a preliminary study. *The Searice Review* **2**(2), 1–8.
205. Sayer, J. (1991) *Rainforest Buffer Zones: Guidelines for Protected Area Managers*, IUCN, Gland, Switzerland.
206. Schall, B.A. (1980) Measurement of gene flow in *Lupinus texensis*. *Nature* **284**, 450–451.
207. Schultes,, R.E. and Hoffmann, H. (1973) *The Botany and Chemistry of Hallucinogens*, C.C. Thomas Publ., London.
208. Seed Savers Network (1997) Internet home page, http://www.om.com.au/seedsave
209. Setyawati, I. (1996) Environmental variability, indigenous knowledge, and the use of rice varieties. *Indigenous Knowledge and Development Monitor* **4**(2), 11–13.
210. Seymour, J. (1996) Hungry for a new revolution. *New Scientist* (30 March), 34–37.
211. Shipek, F. (1996), personal communication to GN.
212. Sjobohm, S. and Singh, A. (1993) An indigenous land tenure system is revived to rehabilitate a protected area: the case of Sariska National Park in Rajasthan. *Forest, Trees and People Newsletter*, No. 22.

213. Smith, N.J.H., Williams, J.T., Plucknett, D.L. and Talbot, J.P. (1992) *Tropical Forests and their Crops*, Comstock Publishing Associates, Ithaca, NY.
214. Snyder, K. (1995), personal communication to GN.
215. Solecki, W.D. (1994) Putting the biosphere reserve concept into practice: some evidence of impacts in rural communities in the United States. *Environmental Conservation* **21**(3), 242–247.
216. Somnasang, P., Rathakette, P. and Rathanapanya, S. (1988) The role of natural foods in northeast Thailand, in *RRA Research Reports*, Khon Khaen University–Ford Foundation Rural Systems Research Project, Khon Khaen, Thailand, pp. 78–102.
217. Stonich, S.C. (1996) Reclaiming the commons: grassroots resistance and retaliation in Honduras. *Cultural Survival Quarterly*, Spring 1996, 31–35.
218. Súzan-Aspiri, H. (1993), personal communication to GN.
219. Súzan-Aspiri, H. (1994) Nurse plant and floral biology of a rare night-blooming cactus, *Penocereus striatus* (Brandegee) F. Buxbaum. *Conservation Biology* **8**(2), 461–470.
220. Tabor, J.A. and Hutchinson, C.F. (1994) Using indigenous knowledge, remote sensing and GIS for sustainable development. *Indigenous Knowledge and Development Monitor* **2**(1), 2–6.
221. Tapia, M. and Rosas, A. (1993) Seed fairs in the Andes: a strategy for local conservation of plant genetic resources, in *Cultivating Knowledge: Genetic Diversity, Farmer Experimentation and Crop Research* (eds W. de Boef, K. Amanor, K. Wellard and A. Bebbington), Intermediate Technology Publications, London.
222. Temple, S.A. (1993), personal communication to JT.
223. Terborgh, J. (1986) Keystone plant resources in the tropical forest, in *Conservation Biology: The Science of Scarcity and Diversity* (ed. M.E. Soule), Sinauer Associates, Sunderland, MA, pp. 330–344.
224. Tewksbury, J.J. and Petrovich, C.A. (1994) The influence of ironwood as a habitat modifier species: a case study on the Sonoran Desert coast of the Sea of Cortez, in *Ironwood: An Ecological and Cultural Keystone of the Sonoran Desert* (eds G.P. Nabhan and J.L. Carr), Occasional Papers in Conservation Biology, No. 1, Conservation International, Washington DC, pp. 29–54.
225. The Nature Conservancy (undated) Vegetation monitoring in a management context: methods for measuring vegetation. Unpublished manuscript, TNC Southeast Regional Office, Chapel Hill, NC.
226. Thiollay, J. (1995) The role of traditional agroforests in the conservation of rain forest bird diversity in Sumatra. *Conservation Biology* **9**(2), 335–353.
227. Toledo, V.M. (1991) in *Biodiversity: Culture, Conservation and Ecodevelopment* (eds M.L. Oldfield and J.B. Alcorn), Westview Press, Boulder, CO.
228. Tuxill, J. (1991) A forest inventory at Bois Diable, Dominica. Unpublished report, prepared for Cottage Forest Industries Ltd and the Caribbean Natural Resources Institute.
229. United Nations Development Program (1995) *Agroecology: Creating the Synergisms for a Sustainable Agriculture*, UNDP, New York.
230. United States National Research Council (1996) *Lost Crops of Africa, Volume I: Grains*, National Academy Press, Washington DC.
231. United States National Research Council (1992a) *Conserving Biodiversity: A Research Agenda for Development Agencies*, National Academy Press, Washington DC.
232. United States National Research Council (1992b) *Sustainable Agriculture and the Environment in the Humid Tropics*, National Academy Press, Washington DC.

233. van Oosterhout, S. (1996) What does *in situ* conservation mean in the life of a small farmer? Examples from Zimbabwe's communal areas, in *Using Diversity: Enhancing and Maintaining Genetic Resources On-Farm* (eds L. Sperling and M. Loevinsohn), International Development Research Centre, New Delhi, India.
234. Vavilov, N.I. (1949) *The Origin, Variation, Immunity and Breeding of Cultivated Plants*, Chronica Botanica, Waltham, MA.
235. Velásquez Runk, J. (1995) Integrating conservation and development: monitoring the ecological impacts of tagua extraction in the Comuna Río Santiago-Cayapas, Esmeraldas Province, Northwest Ecuador. Unpublished Master's thesis, Duke University School of the Environment, Durham, NC.
236. Ventocilla, J., Herrera, H. and Núñez, V. (1995) *Plants and Animals in the Life of the Kuna*, University of Texas Press, Austin, TX.
237. Voeks, R.A. (1996) Tropical forest healers and habitat preference. *Economic Botany* **50**(4), 381–400.
238. Vovides, A.P. and Iglesias, C.G. (1994) An integrated conservation strategy for the cycad *Dioon edule* Lindl. *Biodiversity and Conservation* **3**, 137–141.
239. Wagner, J., Kohaia, V. and Tarihao, F. (1996) The collection of size class structure and recruitment data of *Canarium indicum* by local communities in the Makira Conservation in Development Project Area, Solomon Islands. Unpublished report, Biodiversity Conservation Network, Washington DC.
240. Wells, M.P. and Brandon, K.E. (1993) The principles and practice of buffer zones and local participation in biodiversity conservation. *Ambio* **22**(2–3), 157–162.
241. Whitcombe, J.R. and Joshi, A. (1996a) The impact of farmer participatory research on biodiversity of crops, in *Using Diversity: Enhancing and Maintaining Genetic Resources On-Farm* (eds L. Sperling and M. Loevinsohn), International Development Research Centre, New Delhi, India.
242. Whitcombe, J.R. and Joshi, A. (1996b) Formal participatory approaches for varietal breeding and selection and linkages to the formal seed sector, in *Participatory Plant Breeding* (eds P. Eyzaguirre and M. Iwanaga), International Plant Genetic Resources Institute, Rome.
243. Wilcox, B. (1989) *In situ* conservation of genetic resources, in *The Preservation and Valuation of Biological Resources* (ed. G.H. Orians), University of Washington Press, Seattle, pp. 45–93.
244. Wilkes, H.G. (1993) Conservation of maize crop relatives in Guatemala, in *Perspectives on Biodiversity: Case Studies of Genetic Resource Conservation and Development* (eds C.S. Potter, J.I. Cohen and D. Janczewski), AAAS Press, Washington DC, pp. 75–88.
245. Wilkes, H.G. (1977) Hybridization of maize and teosinte, in Mexico and Guatemala and the improvement of maize. *Economic Botany* **31**, 254–293.
246. Wilkes, H.G. (1972) Genetic erosion of teosinte. *FAO Plant Genetic Resources Newsletter* **28**, 3–10.
247. Wilkes, H.G. (1970) Teosinte introgression in the maize of Nobogame Valley. *Harvard Botanical Museum Leaflet* **22**, 297–311.
248. Wilson, A. (1994) Sacred forests and the elders, in *The Law of the Mother: Protecting Indigenous Peoples in Protected Areas* (ed. E. Kemf), Earthscan Publications, London.
249. Wilson, E.O. (1992) *The Diversity of Life*, Harvard University Press, Cambridge, MA.
250. Worede, M. and Mekbib, H. (1993) Linking genetic resource conservation to farmers in Ethiopia, in *Cultivating Knowledge: Genetic Diversity, Farmer Experimentation and Crop Research* (eds W. de Boef, K. Amanor, K. Wellard and A. Bebbington), Intermediate Technology Publications, London.

251. World Wide Fund for Nature and World Conservation Union (1994) *Centres of Plant Diversity: A Guide and Strategy for their Conservation*, IUCN Publications Unit, Cambridge, UK.
252. Wright, M. (1997), personal communication to JT.
253. Wright, M., Delimini, L., Luhanga, J. *et al.* (1995) *The Quality of Farmer-saved Seed in Ghana, Malawi and Tanzania*, Natural Resources Institute, Chatham, Kent, UK.
254. Yetman, D.A. and Burquez, A. (1996) A tale of two species: speculation on the introduction of *Pachycereus pringlei* in the Sierra Libre, Sonora, México, by *Homo sapiens*. *Desert Plants* **12**(1), 23–32.
255. Zimmerer, K.S. (1996) *Changing Fortunes: Biodiversity and Peasant Livelihood in the Peruvian Andes*, University of California Press, Berkeley, CA.
256. Zimmerer, K.S. (1991) Labor shortages and crop diversity in the southern Peruvian Sierra. *Geographical Review* **81**(4), 414–432.

Index